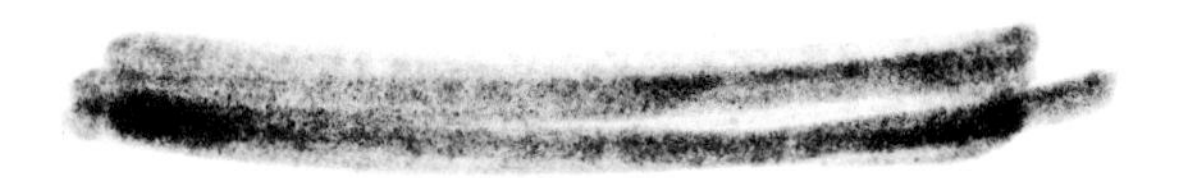

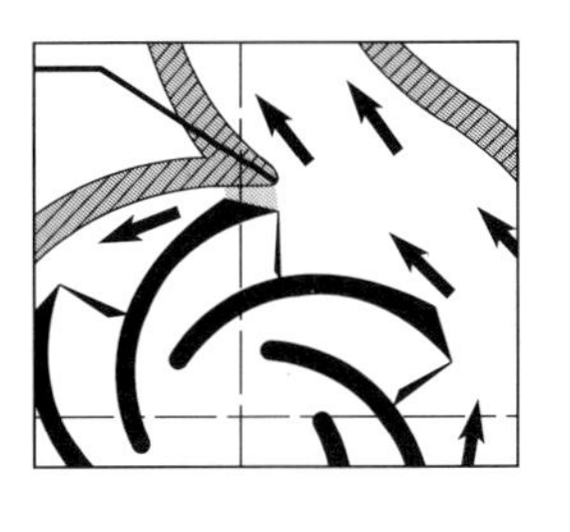

Know and Understand Centrifugal Pumps

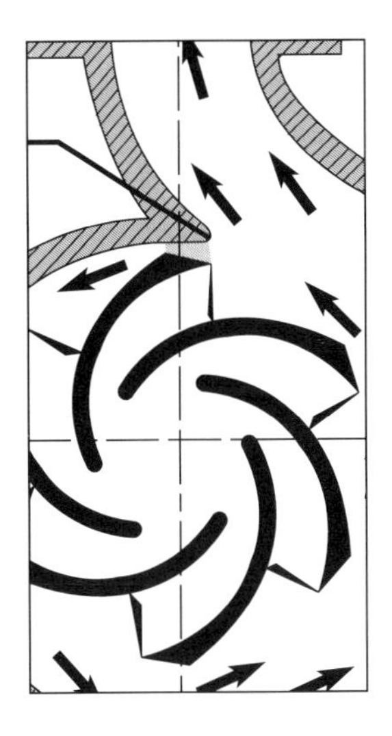

Know and Understand Centrifugal Pumps

by Larry Bachus and Angel Custodio

UK Elsevier Ltd, The Boulevard, Langford Lane, Kidlington, Oxford OX5 1GB, UK
USA Elsevier Inc, 360 Park Avenue South, New York, NY 10010-1710, USA
JAPAN Elsevier Japan, Tsunashima Building Annex, 3-20-12 Yushima, Bunkyo-ku, Tokyo 113, Japan

Published by Elsevier Ltd.

British Library Cataloguing in Publication Data
Bachus, Larry
Know and understand centrifugal pumps
1.Centrifugal pumps
I.Title II.Custodio, Angel
621.6'7

ISBN 1856174093

Published by
Elsevier Advanced Technology,
The Boulevard, Langford Lane, Kidlington, Oxford OX5 1GB, UK
Tel: +44(0) 1865 843000
Fax: +44(0) 1865 843971

Typeset by Land & Unwin (Data Sciences) Ltd, Bugbrooke
Printed and bound in Great Britain by Biddles Ltd, Guildford and King's Lynn

Contents

Prologue

Very few industrial pumps come out of service and go into the maintenance shop because the volute casing or impeller split down the middle, or because the shaft fractured into four pieces. The majority of pumps go into the shop because the bearings or the mechanical seal failed.

Most mechanics spend their time at work time greasing and changing bearings, changing pump packing, and mechanical seals. The mechanical engineers spend their time comparing the various claims of the pump manufacturers, trying desperately to relate the theory learned at the University with the reality of the industrial plant. Purchasing agents have to make costly decisions with inadequate information at their disposal. Process engineers and operators are charged with maintaining and increasing production.

The focus of industrial plant maintenance has always been that the design is correct, and that the operation of the pumps in the system is as it should be. In this book, you will see that in the majority of occasions, this is not true. Most of us in maintenance spend our valuable time, just changing parts, and in the best of cases, performing preventive maintenance, trying to diminish the time required to change those parts.

We almost never stop to consider what is causing the continual failure of this equipment. This book will help you to step away from the fireman approach, of putting out fires and chasing emergencies.

This book is directed toward the understanding of industrial pumps and their systems. It won't be a guide on how to correctly design pumps, nor how to rebuild and repair pumps. There are existing books and courses directed toward those themes. By understanding the real reasons for pump failure, analyzing those failures, and diagnosing pump behavior through interpretation of pressure gauges, you can achieve productive pump operation and contain maintenance costs. This book will serve as a guide to STOP repairing industrial pumps.

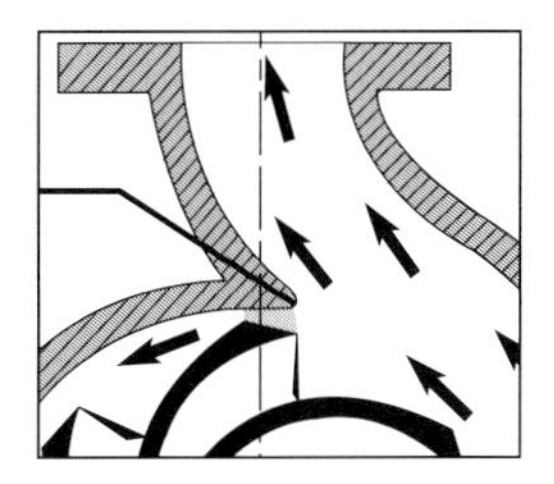

About the Authors

Larry Bachus and Angel Custodio met each other in the early 1990s in Puerto Rico. Larry was working on a pump and seal conversion in a pharmaceutical chemical plant and Angel was installing a computerized preventive maintenance system in the same plant. They had passed each other in the administrative offices at the plant and one day the maintenance engineer introduced them and suggested they work together. They became fast friends and have worked together on numerous projects over the years since, including this book.

Larry Bachus

Larry has almost 30-years experience in maintenance with industrial pumps. His areas of expertise include diagnosing pump problems and seal failures. Larry is highly regarded for his hands-on personalized consulting. He speaks fluent English and Spanish. He has taught pump and seal improvement courses all over the world. His investigations into pump failure have led to inventions, tools and devices used in the chemical process industry. He is an active member of American Society of Mechanical Engineers (ASME) and writes a column called 'The Pump Guy' in *Energy Tech Magazine*.

Angel Custodio

Angel specializes in the installation and implementation of Preventive Maintenance Systems through his consulting engineering company formed in 1987 in Puerto Rico. His installations have given him the opportunity to look into different approaches to hands-on maintenance and operator inspections. Angel conducts seminars on pumps, mechanical components, computerized inventory control and maintenance management. He is also a member of ASME, and the Puerto Rican College of Engineers.

Basic Pump Principles

Introduction

Pumps are used to transfer liquids from low-pressure zones to high-pressure zones:

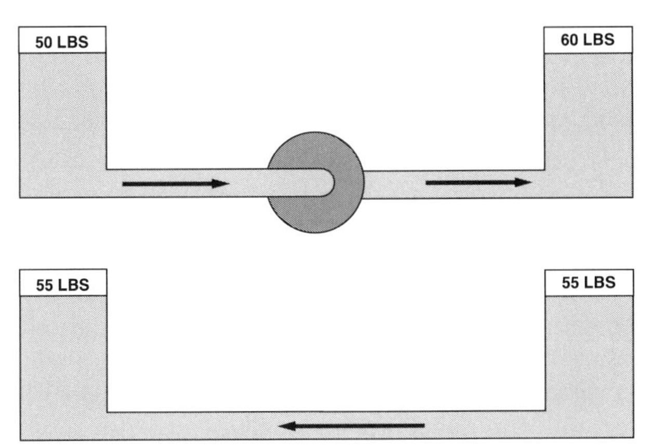

Figure 1–1

Without a pump in this system, the liquid would move in the opposite direction because of the pressure differential.

Pumps are also used to move liquids from a low elevation into a higher elevation, and to move liquids from one place to another. Pumps are also used to accelerate liquids through pipes.

How do pumps work?

The fluid arrives at the pump suction nozzle as it flows through the

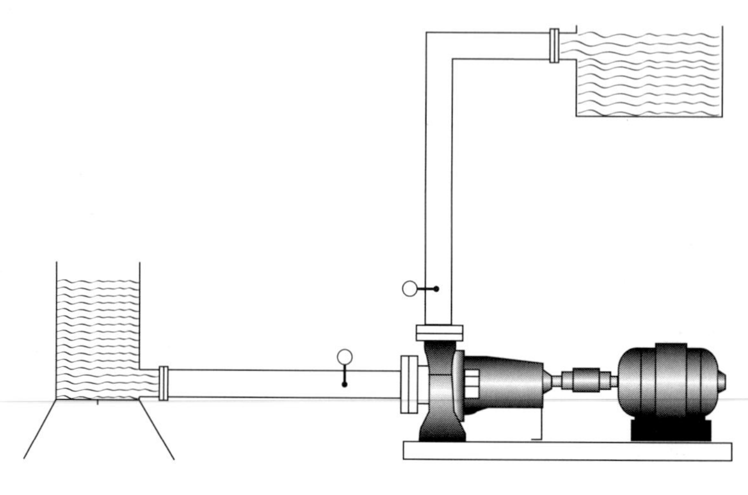

Figure 1-2

suction piping. The fluid must be available to the pump with sufficient energy so that the pump can work with the fluid's energy. The pump cannot suck on or draw the liquid into the pump. The concept of the fluid being available to the pump is discussed in detail in Chapter 2 of this book.

Positive displacement (PD) pumps take the fluid at the suction nozzle and physically capture and contain the fluid in some kind of moveable enclosure. The enclosure may be a housing with a pulsing diaphragm, or between the teeth of rotating gears. There are many designs. The moveable enclosure expands and generates a low pressure zone, to take the fluid into the pump. The captured fluid is physically transported through the pump from the suction nozzle to the discharge nozzle. Inside the pump, the expanded moveable enclosure then contracts or the available space compresses. This generates a zone of high pressure inside the pump, and the fluid is expelled into the discharge piping, prepared to overcome the resistance or pressure in the system. The flow that a PD pump can generate is mostly a function of the size of the pump housing, the speed of the motor or driver, and the tolerances between the parts in relative motion. The pressure or head that a PD pump can develop is mostly a function of the thickness of the casing and the tolerances, and the strength of the pump components.

As the pump performs its duty over time, and fluid passes through the pump, erosion and abrasive action will cause the close tolerance parts to wear. These parts may be piston rings, reciprocating rod seals, a flexing diaphragm, or meshed gear teeth. As these parts wear, the pump will lose its efficiency and ability to pump. These worn parts must be changed with a degree of frequency based on time and the abrasive and lubricating nature of the fluid. Changing these parts should not be

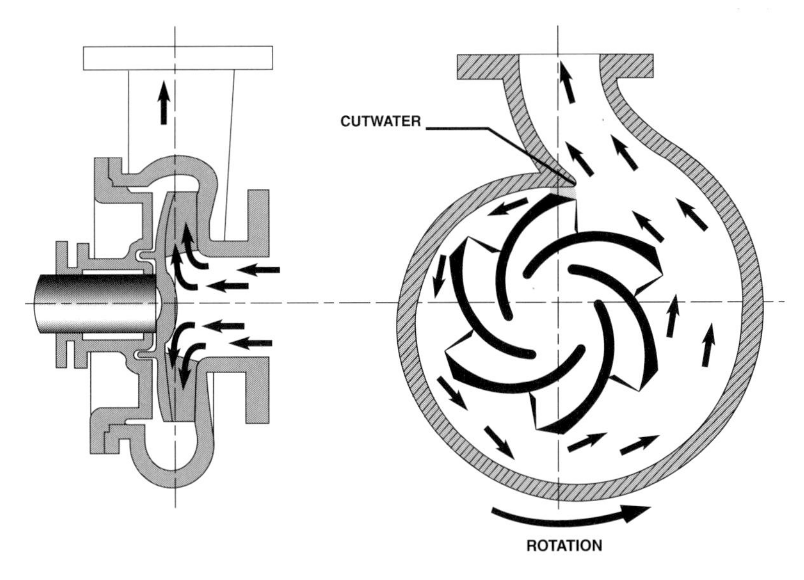

Figure 1-3

viewed as breakdown maintenance. Nothing is broken. This periodic servicing is actually a production function to return the pump to its best or original efficiency.

Centrifugal pumps also require that the fluid be available to the pump's suction nozzle with sufficient energy. Centrifugal pumps cannot suck or draw the liquid into the pump housing. The principal pumping unit of a centrifugal pump is the volute and impeller. (See Figure 1–3).

The impeller is attached to a shaft. The shaft spins and is powered by the motor or driver. We use the term driver because some pumps are attached to pulleys or transmissions. The fluid enters into the eye of the impeller and is trapped between the impeller blades. The impeller blades contain the liquid and impart speed to the liquid as it passes from the impeller eye toward the outside diameter of the impeller. As the fluid accelerates in velocity, a zone of low pressure is created in the eye of the impeller (the Bernoulli Principle, as velocity goes up, pressure goes down). This is another reason the liquid must enter into the pump with sufficient energy.

The liquid leaves the outside diameter of the impeller at a high rate of speed (the speed of the motor) and immediately slams into the internal casing wall of the volute. At this point the liquid's centrifugal velocity comes to an abrupt halt and the velocity is converted into pressure (the Bernoulli Principle in reverse). Because the motor is spinning, there is also rotary velocity. The fluid is conducted from the cutwater around the internal volute housing in an ever-increasing escape channel. As the pathway increases, the rotary velocity decreases and even more energy and pressure is added to the liquid (again Bernoulli's Principle). The liquid leaves the pump at discharge pressure, prepared to overcome the resistance in the system.

The flow from a centrifugal pump is mostly governed by the speed of the driver and the height of the impeller blades. The pressure or head that the pump can generate is mostly governed by the speed of the motor and the diameter of the impeller. Other factors play a lesser role in the pump's flow and pressure, like the number, pitch, and thickness of the impeller blades, the internal clearances, and the presence and condition of the wear bands.

In simple terms, we could say that PD pumps perform work by manipulating the available space inside the pump. Centrifugal pumps perform work by manipulating the velocity of the fluid as it moves through the pump. There is more on this in Chapter 6.

Pressure measurement

Force (F) is equal **to Pressure (P)** multiplied by the **Area (A):**

$\mathbf{F = P \times A}$.

Pressure is equal to the **Force** divided by the **Area:** $\mathbf{P = \frac{F}{A}}$

If we apply pressure to the surface of a liquid, the pressure is transmitted uniformly in all directions across the surface and even through the liquid to the walls and bottom of the vessel containing the liquid (Pascal's Law). This is expressed as pounds per square inch (lbs/in^2, or psi), or kilograms per square centimeter (k/cm^2).

Atmospheric pressure (ATM)

Atmospheric pressure (ATM) is the force exerted by the weight of the atmosphere on a unit of area. **ATM = 14.7 psia** at sea level. As elevation rises above sea level, the atmospheric pressure is less.

Absolute pressure (psia)

Absolute pressure is the pressure measured from a zero pressure reference. Absolute pressure is 14.7 psia at sea level. Compound pressure gauges record absolute pressure.

Gauge pressure (psig)

Gauge pressure is the pressure indicated on a simple pressure gauge. Simple pressure gauges establish an artificial zero reference at atmospheric pressure. The formula is: **psig = psia – ATM**.

Vacuum

The term vacuum is used to express pressures less than atmospheric pressure (sometimes represented as a negative psi on pressure gauges). Another scale frequently used is 'inches of mercury'. The conversion is: **14.7 psia = 29.92″** Hg. Another scale gaining in popularity is the kilopascal (Kp) scale. **14.7 psia = 100 Kp**

AUTHOR'S NOTE

Note that there are many ways to express vacuum. Simple gauges record vacuum as a negative psig. Compound gauges record vacuum as a positive psia. The weatherman uses inches of mercury in the daily forecast, and millibars (1000 millibars is atmospheric pressure) to express the low-pressure zone in the eye of a hurricane. Boiler operators use water column inches and millimeters of mercury to express vacuum.

Pump manufacturers express vacuum in aspirated feet of water in a vertical column **(0 psia = –33.9 feet of water)**. The pharmaceutical and chemical industry uses 'Pascals' **(100,000 Pascals = atmospheric pressure)** and the term TORR. This conglomeration of values and conversion rates causes confusion. In order to understand pumps, it's best to think of vacuum as a positive number less than 14.7 psi. In our experience, we've found that considering vacuum in this form aids the understanding of net positive suction head (NPSH), cavitation, suction specific speed (Nss), and the ability of pumps to suck-up (actually pumps don't suck, but this will do for now) fluid from below. Remember that vacuum is the absence of atmospheric pressure, but it is not a negative number.

Pump head

The term 'pump head' represents the net work performed on the liquid by the pump. It is composed of four parts. They are: the static head (Hs), or elevation; the pressure head (Hp) or the pressures to be overcome; the friction head (Hf) and velocity head (Hf), which are frictions and other resistances in the piping system. These heads are discussed in Chapter 8. The head formula is the following:

$$H = \frac{P}{D}$$

Where: **H** = head **P** = psi **d** = density

Pressure can be converted into head with the following equation:

$$\textit{Head ft.} = \frac{2.31 \times \textit{Pressure psi}}{\textit{sp.gr.}}$$

Where: **H** = head in feet **psi** = pressure in pounds per square inch
2.31= conversion factor **sp. gr.** = specific gravity

Head converts to pressure with the following formula:

$$\textit{Pressure psi} = \frac{\textit{Head ft.} \times \textit{sp.gr.}}{2.31}$$

Specific gravity

Specific gravity is the comparison of the density of a liquid with the density of water. With pumps, it is used to convert head into pressure. The specific gravity formula is:

$$\textit{Sp.Gr.} = \frac{\textit{Density Liquid}}{\textit{Density Water}}$$

The standard for water is 60°F at sea level.

Water is designated a specific gravity of 1.0. Another liquid is either heavier (denser) or lighter than water. The volume is not important as long as we compare equal volumes. The specific gravity affects the pressure in relation to the head, and it affects the horsepower consumed by the pump with respect to pressure and flow. We'll study this in depth later.

Pressure measurement

Pressure exists in our daily lives. At sea level the atmospheric pressure is 14.7 psia. This is the pressure exerted on us by the air we breathe. If we should remove all the air, then the pressure would be zero.

We're more concerned with pressures above atmospheric pressure. For example, a flat tire on a car still has 14.7 pounds of pressure inside it. We would consider this to be a flat tire because the pressure outside the tire is equal to the pressure inside the tire. We would say the tire has no pressure because it would not be inflated and could not support the weight of the car.

What is more important to us is the differential pressure inside the tire compared to outside the tire (atmospheric pressure). For reasons such as these, the world has adopted a second and artificial zero, at atmospheric pressure as a reference point. This is why a simple pressure gauge will read zero at atmospheric pressure.

Because simple pressure gauges are made with an artificial zero at atmospheric pressure, this is why the term psig exists, meaning pounds per square inch gauge. As mentioned earlier, the psig is equal to the absolute pressure minus the atmospheric pressure.

$$\mathbf{Psig = Psia - ATM}$$

Pressures less than atmospheric are recorded as negative pressures (–psi) on a simple pressure gauge.

Technically speaking, negative pressures don't exist. Pressure is only a positive force and it is either present or absent.

Pressures inside the pump

Suction pressure

Suction pressure is the pressure at the pump's suction nozzle as measured on a gauge. The suction pressure is probably the most important pressure inside the pump. All the pump's production is based on the suction pressure. The pump takes suction pressure and converts it into discharge pressure. If the suction pressure is inadequate, it leads to cavitation. Because of this, all pumps need a gauge at the suction nozzle to measure the pressure entering the pump.

Discharge pressure

This is the pressure at the pump discharge nozzle as measured by a gauge. It is equal to the suction pressure plus the total pressure developed by the pump.

Seal chamber pressure

This is the pressure measured in the stuffing box or seal chamber. This is the pressure to be sealed by the mechanical seal or packing. The seal chamber pressure must be within the limits of the mechanical seal. This

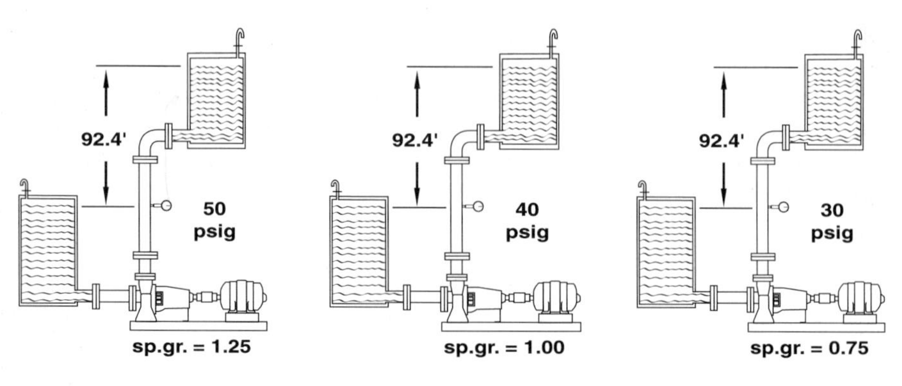

Figure 1–4

pressure is very important with double mechanical seals, because it governs the pressure setting of the barrier fluid.

Head versus pressure

Figures 1–4 and 1–5 show the relationship between head and pressure in a centrifugal pump moving liquids with different specific gravities. There is more on this in Chapter 7.

The above graphic shows three identical pumps, each designed to develop 92.4 feet of head. When they pump liquids of different specific gravities, the heads remain the same, but the pressures vary in proportion to the specific gravity.

In the graphic below (Figure 1–5), these three pumps are developing the same discharge pressure. In this case they develop different heads inversely proportional to the specific gravity of the fluids.

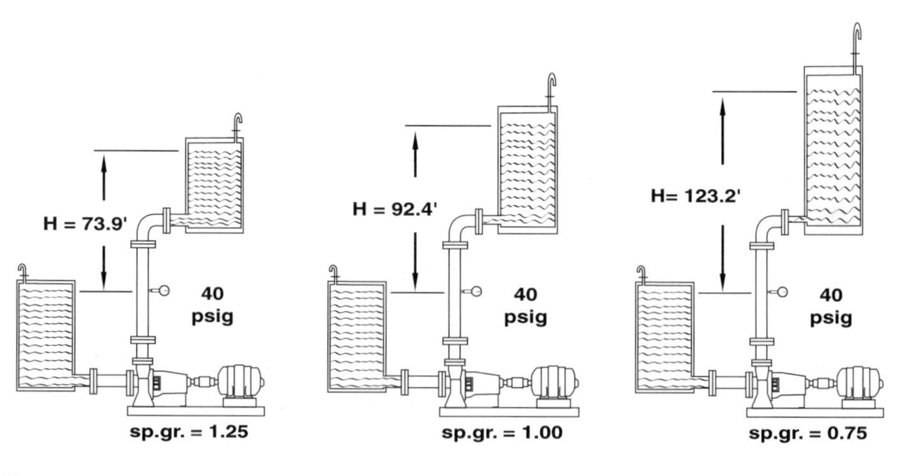

Figure 1–5

AUTHOR' NOTE

The concept of Head versus Pressure causes confusion between maintenance people and the pump manufacturer. The maintenance technician reads his gauges recording pressure in psi, and the pump manufacturer uses the term head. The term head is the constant for the manufacturer. A pump that generates 90 feet of head can elevate water, gasoline, caustic soda, and any liquid to a height of 90 feet. The manufacturer doesn't know the ultimate service of the pump when he manufactures it. He only knows that his pump will develop 90 feet of head. The psi reading is a function of the conversion factor 2.31 and also the specific gravity. This is why you cannot specify a pump by the psi. If the maintenance engineer or mechanic wants to have an intelligent conversation with the pump manufacturer, he must understand and use the concept of 'head'. This is also the reason that too many pumps are sold without adequate gauges. It's somewhat like selling a car without a dashboard. There's more information on this in Chapters 7 and 8.

Given the following information:

sp. gr. of water = 1.0

sp. gr. of gasoline = 0.70

sp. gr. of concentrated sulfuric acid = 2.00

sp.gr. of sea water = 1.03

A pump capable of generating 125 feet of head would provide the following pressures:

Pressure = (Head ft. × sp.gr.) / 2.31

Water: $$P = \frac{1.25 \times 1.0}{2.31} = 54.1 \text{ psig}$$

Gasoline: $$P = \frac{1.25 \times 0.7}{2.31} = 37.8 \text{ psig}$$

Conc. Sulfuric Acid: $$P = \frac{1.25 \times 2.0}{2.31} = 108.2 \text{ psig}$$

Sea Water $$P = \frac{1.25 \times 1.03}{2.31} = 55.3 \text{ psig}$$

This pump (Figure 1–6) is raising the liquid from the level in the suction vessel to the level in the discharge vessel. This distance is called the **Total Head**.

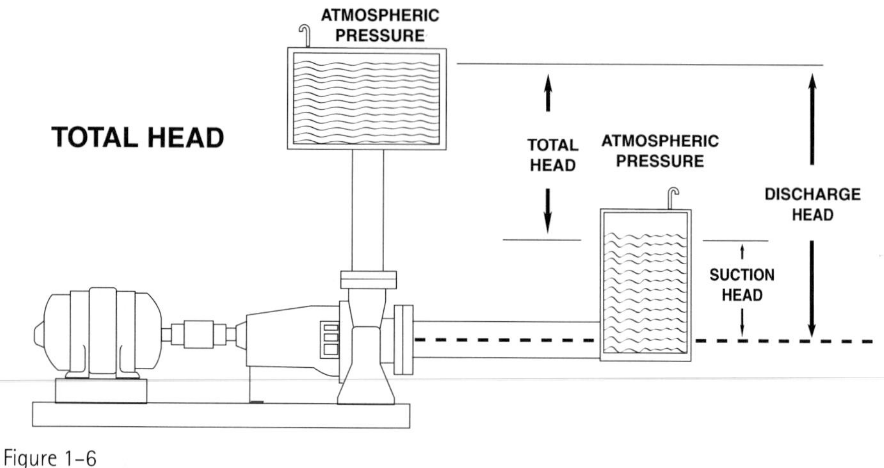

Figure 1-6

The total head is:

- The work of the pump.
- The measure of the pump's ability to raise the liquid to a given height.
- The measure of the pump's ability to develop a given discharge pressure.
- The discharge elevation minus the suction elevation.
- The discharge head minus the suction head.
- The discharge head plus the suction lift.
- The discharge absolute pressure reading minus the suction absolute pressure reading.

Suction head

The suction head is the available head at the suction nozzle of the pump.

Discharge head

The discharge head is the vertical distance from the centerline of the pump (this would be the shaft on a horizontal pump) to the level in the discharge vessel.

Suction lift

Suction lift is negative suction head. It exists when the liquid level in the suction vessel is below the centerline of the pump. The pump must aspirate the liquid up from the suction vessel into the pump and then

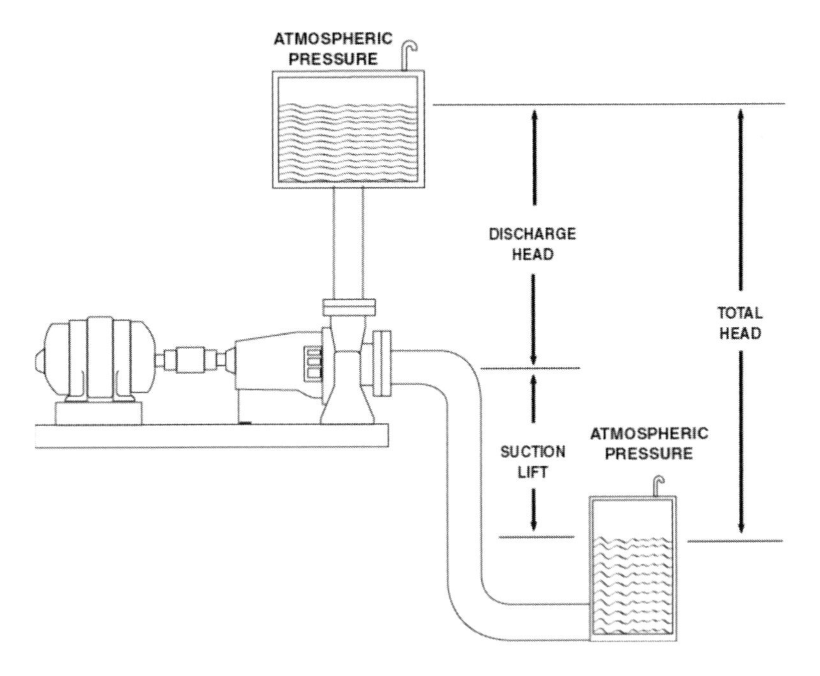

Figure 1–7

push the liquid up into the discharge vessel. This pump (Figure 1–7) is said to be in suction lift.

In this case, the pump must **aspirate** or lift the liquid up from the suction vessel into the pump and then **push** the liquid up into the discharge vessel. In this case the **total head** is the discharge head *plus* the suction lift. In all cases the **total head** is the work being performed by the pump.

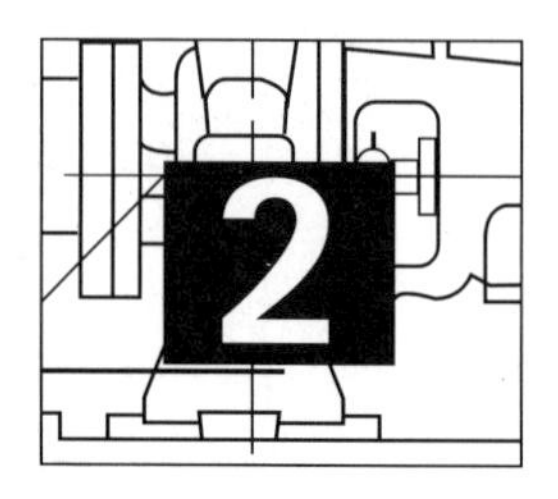

NPSH, Net Positive Suction Head

Introduction

When someone turns on an electric light, the natural tendency is to look toward the light and consider the shine. We tend not to think about the electric wires and the current running through the light bulb. Equally, when someone starts an industrial pump, the tendency is to look toward the discharge piping and consider the pressure and flow. We tend not to think about the suction piping, or the liquid coming into the eye of the impeller. We need to emphasize the necessity to consider what's happening in the suction of the pump. This area is the source of problems, and probably is responsible for about 40% of all pumps going into the shop today.

This chapter is dedicated to NPSH, Net Positive Suction Head. NPSH is what the pump needs, the minimum requirement to perform its duties. Therefore, NPSH is what happens in the suction side of the pump, including what goes on in the eye of the impeller. NPSH takes into consideration the suction piping and connections, the elevation and absolute pressure of the fluid in the suction piping, the velocity of the fluid and the temperature. For the moment we can say that some of these factors add energy to the fluid as it moves into the pump, and others subtract energy from the fluid. There must be sufficient energy in the fluid for the impeller to convert this energy into pressure and flow. If the energy is inadequate we say that the pump suffers inadequate NPSH.

In simple terms we could say that NPSH is the reason that the suction nozzle is generally larger than the discharge nozzle. If there is more liquid leaving the pump faster than the liquid can enter into the pump, then the pump is being starved of liquid.

AUTHOR'S NOTE

Think about it this way. When we see a magician pulling a rabbit out of a hat, in all probability there's a rabbit hidden in a secret compartment inside the top hat, or the rabbit is hidden in the magician's coat sleeve. The rabbit does not appear spontaneously. Isn't it interesting that magicians all wear long sleeved topcoats? They always reach into a 'top hat' for the rabbit. When I see a magician pull a rhinoceros from a frisbee, then maybe I'll believe in magic. There is illusion, but there is no magic. Likewise with a pump, the energy must be in the fluid for the impeller to convert it.

Equally, if your body requires more oxygen than the available oxygen in the atmosphere, then you would be asphyxiated. There must be more oxygen available in the air than the oxygen you consume.

To express the quantity of energy available in the liquid entering into the pump, the unit of measure for NPSH is feet of head or elevation in the pump suction. The pump has its NPSHr, or Net Positive Suction Head Required. The system, meaning all pipe, tanks and connections on the suction side of the pump has the NPSHa, or the Net Positive Suction Head Available. There should always be more NPSHa in the system than the NPSHr of the pump. Let's look at them, beginning with what the pump requires:

Definition of NPSHr (required)

It is the energy in the liquid required to overcome the friction losses from the suction nozzle to the eye of the impeller without causing vaporization. It is a characteristic of the pump and is indicated on the pump's curve. It varies by design, size, and the operating conditions. It is determined by a lift test, producing a negative pressure in inches of mercury and converted into feet of required NPSH.

AUTHOR'S NOTE

An easy way to understand NPSHr is to call it the minimum suction pressure necessary to keep the pumped fluid in a liquid state.

According to the Standards of the Hydraulic Institute, a suction lift test is performed on the pump and the pressure in the suction vessel is lowered to the point where the pump suffers a 3% loss in total head. This point is called the NPSHr of the pump. Some pump manufacturers perform a similar test by closing a suction valve on a test pump and other manufacturers lower the suction elevation.

The definition of NPSHr may change in the future. A pump is in a definite state of cavitation with the 3% total head loss definition. Many pump users want a more explicit definition of NPSHr, and higher NPSHa safety margins to avoid inadequate NPSHa and cavitation altogether.

The pump manufacturers publish the NPSHr values on their pump curves. We're saying that the NPSH reading is one of the components of your pump curves. We'll see this in Chapter 7 on Pump Curves. If you want to know the NPSHr of your pump, the easiest method is to read it on your pump curve. It's a number that changes normally with a change in flow. When the NPSHr is mentioned in pump literature, it is normally the value at the best efficiency point. Then, you'll be interested in knowing exactly where your pump is operating on its curve.

If you don't have your pump curve, you can determine the NPSH of your pump with the following formula:

$$NPSHr = ATM + Pgs + Hv - Hvp$$

Where: **ATM** = the atmospheric pressure at the elevation of the installation expressed in feet of head.
Pgs = the suction pressure gauge reading taken at the pump centerline and converted into feet of head.
Hv = Velocity Head = $V^2/2g$ where: V = the velocity of the fluid moving through the pipes measured in feet per second, and 'g' = the acceleration of gravity (32.16 ft/sec).
Hvp = the vapor pressure of the fluid expressed in feet of head. The vapor pressure is tied to the fluid temperature.

The easiest thing to do is to get the pump curve from the manufacturer because it has the NPSHr listed at different flows. Nowadays, you can get the pump curve on the Internet with an e-mail to the manufacturer, you can send a fax, or request the curve in the mail or with a local call to the pump representative or distributor. If you wanted to verify the NPSHr on your pump, you'll need a complete set of instrumentation: a barometer gauge, compound pressure gauges corrected to the centerline of the pump, a flow meter, a velocity meter, and a thermometer. Definitely, it's easier to get the curve from your supplier.

Definition NPSHa (available)

This is the energy in the fluid at the suction connection of the pump over and above the liquid's vapor pressure. It is a characteristic of the system and we say that the NPSHa should be greater than the NPSHr (NPSHa > NPSHr).

As a general guide the NPSHa should be a minimum 10% above the NPSHr or 3 feet above the NPSHr, whichever is greater. Other books and experts indicate that the NPSHa should be 50% greater than the NPSHr, to avoid incipient cavitation. Again, be prepared for stricter definitions to NPSHr and higher safety margins on NPSHa.

The NPSHa is in the system. The formula is:

$$NPSHa = Ha + Hs - Hvp - Hf - Hi$$

Where: **Ha** = Atmospheric head (14.7 psi × 2.31) = 33.9 ft. at sea level. See Properties of Water I in this chapter that considers atmospheric pressure at different elevations above sea level.
Hs = Static head in feet (positive or negative) of the fluid level in the suction vessel to the pump centerline.
Hvp = the Vapor head of the fluid expressed in feet. It is a function of the temperature of the liquid. See Properties of Water II in this chapter.
Hf = Friction head or friction losses expressed in feet in the suction piping and connections.
Hi = Inlet head, or the losses expressed in feet that occur in the suction throat of the pump up to and including the eye of the impeller. These losses would not be registered on a suction pressure gauge. They could be insignificant, or as high as 2 feet. Some pump manufacturers factor them into their new pumps, and others don't. Also, changes occur in maintenance that may alter the Hi. If you don't know the Hi, call it a safety factor of 2 feet.

By observing the system, you can calculate the NPSHa within a one or two point margin. The main idea is to be sure the NPSHa is greater than the NPSHr of the pump. Remember that the NPSHa only deals with the suction side of the pump. Let's go back to that formula:

$$NPSHa = Ha + Hs - Hvp - Hf - Hi$$

1. To determine the Ha, atmospheric head, you only need observe the vessel being drained by the pump. Is it an opened, or vented atmospheric vessel? Or is it a closed and sealed vessel? If the vessel is open, then we begin with the atmospheric pressure expressed in feet, which is 33.9 feet at sea level. The altitude is important. The atmospheric pressure adds energy to the fluid as it enters the pump. For closed un-pressurized vessels the Ha is equal to the Hvp and they cancel themselves. For a closed pressurized vessel remember that every 10 psia of pressure on a vessel above the vapor head of the fluid will add 23.1 feet of Ha. To the Ha, we add the Hs.
2. The Hs, static head, is the static height in feet observed from the level in the vessel to be drained to the centerline of the pump. If the

Properties of water I – Atmospheric and barometric pressure readings at different altitudes

Altitude		Barometric pressure		Atmospheric pressure		Boiling point of water °F
Feet	Meters	In. Hg.	mm. Hg.	Psia	Feet water	
-1000	-304.8	31.0	788	15.2	35.2	213.8
-500	-152.4	30.5	775	15.0	34.6	212.9
0	0.0	29.9	760	14.7	33.9	212.0
+500	+152.4	29.4	747	14.4	33.3	211.1
+1000	304.8	28.9	734	14.2	32.8	210.2
1500	457.2	28.3	719	13.9	32.1	209.3
2000	609.6	27.8	706	13.7	31.5	208.4
2500	762.0	27.3	694	13.4	31.0	207.4
3000	914.4	26.8	681	13.2	30.4	206.5
3500	1066.8	26.3	668	12.9	29.8	205.6
4000	1219.2	25.8	655	12.7	29.2	204.7
4500	1371.6	25.4	645	12.4	28.8	203.8
5000	1524.0	24.9	633	12.2	28.2	202.9
5500	1676.4	24.4	620	12.0	27.6	201.9
6000	1828.8	24.0	610	11.8	27.2	201.0
6500	1981.2	23.5	597	11.5	26.7	200.1
7000	2133.6	23.1	587	11.3	26.2	199.2
7500	2286.0	22.7	577	11.1	25.7	198.3
8000	2438.4	22.2	564	10.9	25.2	197.4
8500	2590.8	21.8	554	10.7	24.7	196.5
9000	2743.2	21.4	544	10.5	24.3	195.5
9500	2895.6	21.0	533	10.3	23.8	194.6
10000	3048.0	20.6	523	10.1	23.4	193.7
15000	4572.0	16.9	429	8.3	19.2	184.0

level in the tank is 10 feet above the pump then the Hs is 10. A positive elevation adds energy to the fluid and a negative elevation (suction lift condition) subtracts energy from the fluid. To the sum of the Ha and Hs, we subtract the Hvp.

3. The Hvp, vapor head, is calculated by observing the fluid temperature, and then consulting the water properties graph in this chapter. Let's say we're pumping water at 50° F (10° C). The Hvp is 0.411 feet. If the water is 212° F (100° C) then the Hvp is 35.35 feet. The vapor head is subtracted because it robs energy from the fluid in the suction pipe. Remember that as the temperature rises, more energy is being robbed from the fluid. Next, we must subtract the Hf.

Properties of water II – Vapor Pressure

Temp. °F	Temp. °C	Specific Gravity 60 °F	Density	Vapor Pres. psi	Vapor Pressure* Feet Abs.
32	0	1.002	62.42	0.0885	0.204
40	4.4	1.001	62.42	0.1217	0.281
45	7.2	1.001	62.40	0.1475	0.34
50	10	1.001	62.38	0.1781	0.411
55	12.8	1.000	62.36	0.2141	0.494
60	15.6	1.000	62.34	0.2563	0.591
65	18.3	0.999	62.31	0.3056	0.706
70	21.1	0.999	62.27	0.6331	0.839
75	23.9	0.998	62.24	0.4298	0.994
80	26.7	0.998	62.19	0.5069	1.172
85	29.4	0.997	62.16	0.5959	1.379
90	32.2	0.996	62.11	0.6982	1.617
95	35.0	0.995	62.06	0.8153	1.890
100	37.8	0.994	62.00	0.9492	2.203
110	43.3	0.992	61.84	1.275	2.965
120	48.9	0.990	61.73	1.692	3.943
130	54.4	0.987	61.54	2.223	5.196
140	60.0	0.985	61.39	2.889	6.766
150	65.6	0.982	61.20	3.718	8.735
160	71.1	0.979	61.01	4.741	11.172
170	76.7	0.975	60.79	5.992	14.178
180	82.2	0.972	60.57	7.510	17.825
190	87.8	0.968	60.35	9.339	22.257
200	93.3	0.964	60.13	11.526	27.584
212	100.0	0.959	59.81	14.696	35.353
220	104.4	0.956	59.63	17.186	41.343
240	115.6	0.948	59.10	24.97	60.77
260	126.7	0.939	58.51	35.43	87.05
280	137.8	0.929	58.00	49.20	122.18
300	148.9	0.919	57.31	67.01	168.22
320	160.0	0.909	56.66	89.66	227.55
340	171.1	0.898	55.96	118.01	303.17
360	182.2	0.886	55.22	153.04	398.49
380	193.3	0.874	54.47	195.77	516.75

4. The Hf, friction head, can be calculated, approximated, or measured. The friction head can be calculated with the friction tables for pipe and fittings. You can consult the Hazen Williams formula, or the Darcy Weisbach formula mentioned in Chapter 8 of this book. The friction head can be measured with gauges using the

Bachus Custodio formula explained in Chapter 8. In most cases, the pump is relatively close to the vessel being drained by the pump. In this case the Hf is probably negligible. Hf is subtracted because friction in the suction pipe robs energy from the fluid as it approaches the pump.

5. The Hi, inlet head, is simply a safety factor of 2 feet. Some pumps have an insignificant Hi. Other pumps have inlet losses approaching 2 feet. The Hi is losses to the fluid after it passes the suction pressure gauge and goes into the impeller eye. In a maintenance function, you can't be precise about what's happening to the fluid in this part of the pump. Just call it 2 feet.

Now let's apply the hints and the formula to the following system figures and we can determine the NPSHa within one or two points. The important thing is that the NPSHa of the system is greater than the NPSHr of the pump. If the NPSHa should be inadequate, the pump is being starved, becomes unstable and cannot perform its duties. The inadequate NPSHa may lead to cavitation.

Remember that **NPSHa > NPSHr**

This open system pumping water is at sea level (Figure 2–1). Therefore the Ha is 33.9 feet. The level in the tank is 15 feet above the pump centerline, so the Hs_1 is 15 feet. The friction losses in the suction piping give us 2 feet. The water is 70° F so the Hvp is 0.839. The Hi is a safety factor of 2 feet.

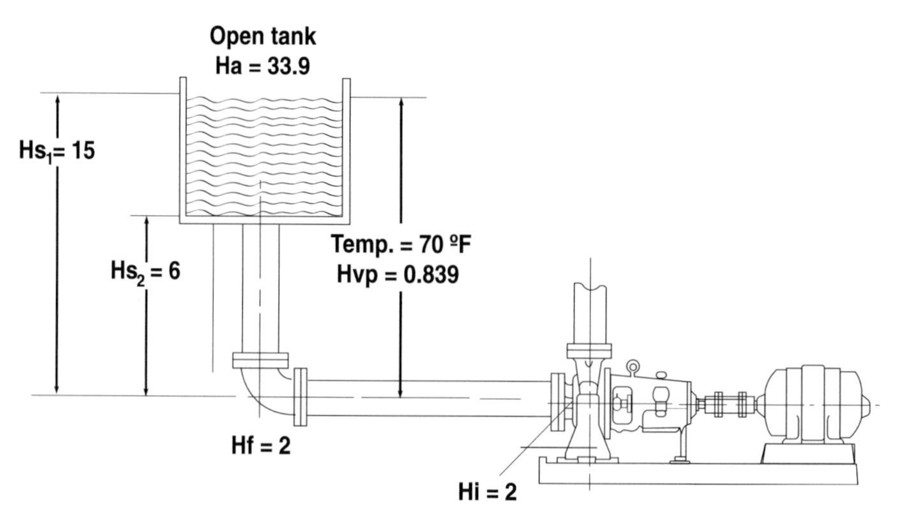

Figure 2–1

NPSHa = Ha + Hs_1 –Hvp – Hf – Hi
NPSHa = 33.9 + 15.0 –0.839 – 2.0 – 2.0
NPSHa = 44.061 feet

The curve of the pump in this service should show an NPSHr of less than 44 ft at the duty point. And the purpose of this pump is to drain this tank, lowering its level. If we don't want inadequate NPSHa and the possible resulting cavitation to start during the process we should consider a second Hs_2 with the tank empty. The other factors remain the same. At the end of the process, we have:

NPSHa = Ha + Hs_2 – Hvp – Hf – Hi
NPSHa = 33.9 + 6.0 – 0.839 – 2.0 – 2.0
NPSHa = 35.061 feet

To avoid stress from inadequate NPSHa during the draining process, we should consult the pump curve and be sure that the NPSHr is less than 35 ft at the duty point.

Now let's consider Figure 2–2. This is a pump in suction lift draining an opened tank that's 8 feet below the pump centerline. This pump is installed high on a mountain at 7,000 feet above sea level. The Ha is 26.2 feet. The Hs_1 is –8.0 feet. The water temperature is 50° F, so the Hvp is 0.411. The Hf is 1 foot and the Hi is 2.0. According to the information:

NPSHa = Ha + Hs_1 – Hvp – Hf – Hi
NPSHa = 26.2 + (–8.0) – 0.411 – 1.0 – 2.0
NPSHa = 14.8 feet

The curve of the pump in this service should show a NPSHr of less than 14 feet at the duty point. The purpose of this pump is to drain this tank down to 14 feet below the pump without cavitating. Let's consider a second static head, Hs^2, of –14 feet. The other factors would remain the same:

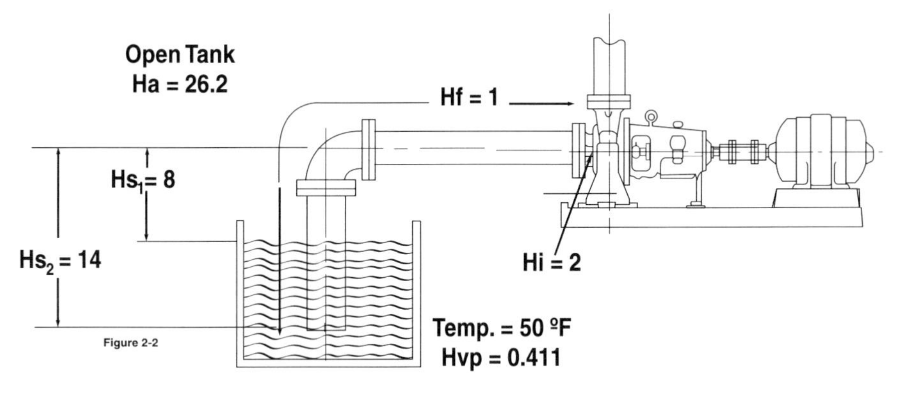

Figure 2–2

NPSHa = Ha + Hs_2 –Hvp – Hf – Hi
NPSHa = 26.2 + (–14.0) – 0.411 – 1.0 – 2.0
NPSHa = 8.8 feet

To avoid problems with this pump during the process, be sure the pump curve indicates NPSHr less than 8 ft at the duty point.

Many processes use sealed tanks and reactor vessels. For example, in a milk processing plant or a pharmaceutical plant, it's necessary to prevent outside air from contaminating the sterile product. In a beer brewery, you can't let the gas and carbonization escape from the process. In a closed un-pressurized vessel, the Ha is equal to the Hvp. And because the Ha adds energy and the Hvp subtracts energy, they cancel themselves. The formula is simpler:

NPSHa = Hs – Hf – Hi

The level in this sealed tank is 12 feet above the pump (Figure 2–3). The Hs_1 is 12 feet. The purpose of this pump is to drain this tank to a level 6 feet above the pump, so the Hs_2 is 6 feet. The Hf is 1.5 feet and the Hi is 2 feet.

NPSHa = Hs_1 – Hf – Hi
NPSHa = 12.0 – 1.5 – 2.0
NPSHa = 8.5

The curve of the pump that drains this tank should register an NPSHr

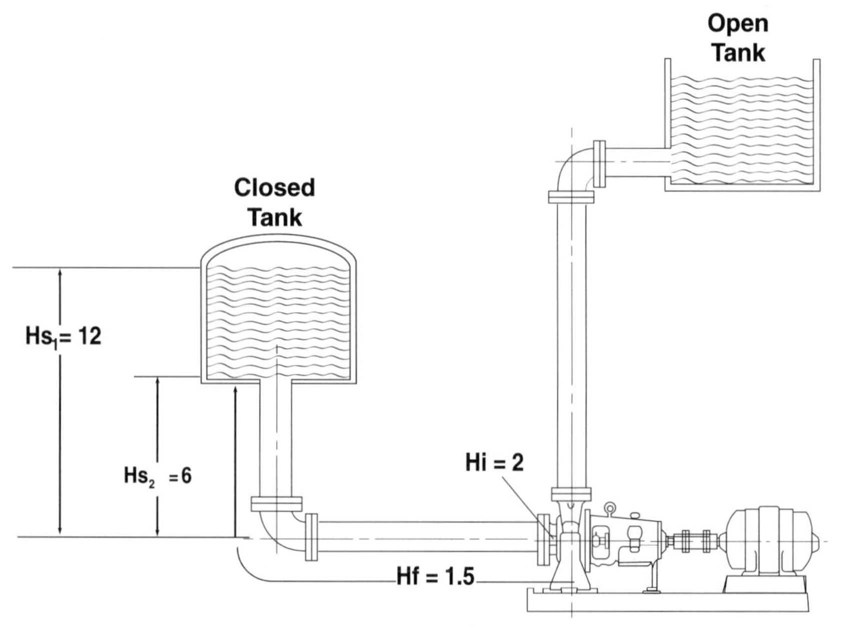

Figure 2–3

of less than 8 feet at the duty point. And, to be sure that problems don't arise during the process, we could calculate the NPSHa at the end of the process:

$NPSHa = Hs_2 - Hf - Hi$
$NPSHa = 6.0 - 1.5 - 2.0$
$NPSHa = 2.5$ feet

Now, it's one thing to say to use a pump with an NPSHr less than 2 feet. It's another thing to find a pump with this design parameter, that at the same time complies with the demands of the operation. Perhaps it will be necessary to modify the system to increase the Hs_2, reduce the Hf, or modify the pump to reduce the Hi. Other possible options are:

1. Pressurize the tank with air or a gas compatible with the liquid and process.
2. Turn off the pump and drain the tank by gravity.
3. Install a small booster pump that feeds the principal pump.
4. Operate the pump at a slower speed.
5. Survive the cavitation. (There's a discussion on this later in the book.)

As we've said numerous times before in this chapter, the important thing is that the NPSHa of the system is above the NPSHr of the pump by a sufficient amount to avoid stress and possible cavitation. If the NPSHa should be inadequate, there are ways to elevate it. Remember from the formula that five elements compose the NPSHa. Two of those elements, the Ha and the Hs, add energy to the fluid. And three elements, the Hvp, the Hf, and the Hi, subtract energy from the fluid. We must either increase the elements that add energy, or decrease the elements that subtract energy. To increase the NPSHa:

1. Raise the level in the tank if possible. This adds **Hs**.
2. Elevate the tank maybe with stilts. This adds **Hs**.
3. Maybe you can lower the pump. For example in many thermoelectric plants, the fuel oil pumps (#6 bunker fuel) are in a pit. This would permit draining the tanks down to the ground and still maintain 15 or 20 feet of NPSHa on the fuel oil pumps. This adds **Hs**.
4. Pressurize the tank if possible. This adds **Ha**.
5. Reduce the drag (Hf) in the suction piping. Change to larger diameter suction piping, or reduce the pipe schedule (change from 'schedule 40 pipe' to 'schedule 20 pipe' on the suction side). Investigate changing the pipe material. For example PVC pipe, and food grade Stainless, is rather slick on the ID. This reduces **Hf**.

6. Reduce the losses (Hf) of the connections and fittings in the suction piping. For wheel actuation valves, maybe globe valves could be converted into gate valves. For quarter turn valves, butterfly valves could be replaced with ball valves. A totally open butterfly valve still has the post and wings in the flow path. Maybe convert short radius elbows into long radius elbows. If you had two or three consecutive elbows, maybe you could use a flexible 'S' connection. This reduces **Hf**.
7. Eliminate some elbows. If the suction piping has multiple elbows, you can bet that some of those elbows are canceling themselves, and are not needed. This reduces **Hf**.
8. Lower the temperature of the fluid in the suction. This reduces the **Hvp**.

If you cannot increase the NPSHa of the system, maybe you could reduce the NPSHr of the pump, by:

1. Change to a pump with a larger suction diameter. For example, convert a 1 × 2 × 8 pump, into a 2 × 3 × 8 pump. The larger pump would have a reduced NPSHr. You need to keep the same impeller diameter (8 inch) to maintain the discharge head and pressures, but you would be converting the 2 inch suction nozzle into a 3 inch suction nozzle. This would reduce the fluid velocity entering into the pump, and therefore the **Hf** and **Hi**.
2. Install a small booster pump into the suction piping. The booster pump would have a reduced NPSHr for the system feeding it, and the discharge head of the booster pump would increase the **Ha** to the primary pump.
3. Increase the diameter of the eye of enclosed impellers. This reduces **Hi**.
4. Ream out and polish the suction throat and pathway to the impeller. This is normally the roughest casting inside the pump. Center the suction nozzle on a lathe and open the diameter of the pathway toward the impeller. This lowers the existing NPSHr of your pump, reducing the **Hi**.
5. Use an impeller inducer. An impeller inducer looks like a corkscrew device that fits onto the center hub of the primary impeller and extends down the suction throat of the pump. It is actually a small axial flow impeller that accelerates the fluid toward the primary impeller from further down the suction throat of the pump. Some inducers bolt onto the impeller and others are cast into the main impeller. The inducer has a low NPSHr for the system feeding it, and it increases the **Ha** to the primary impeller.

6. Convert to a pump with a double suction impeller. Double suction impeller pumps are for low NPSH applications.
7. Use two smaller pumps in parallel.
8. Use a larger/slower pump.

Inadequate NPSHa causes stress, vibration and maintenance on pumps because there is not enough energy in the fluid for the pump to perform its work. As you can see from the previous pages, the problems lie in system design and proper operating principles. When the NPSHa is below the NPSHr of the pump, the conditions are favorable for the pump to go into cavitation. Cavitation is the next chapter.

Cavitation

Introduction

It is important to clarify that the pump does not cavitate, although people in the industry tend to say that the pump is cavitating. It is more correct to say that the pump is in cavitation or the pump is suffering cavitation. In reality it is the system that cavitates the pump, because the system controls the pump.

Inadequate NPSHa establishes favorable conditions for cavitation in the pump. If the pressure in the eye of the impeller falls below the vapor pressure of the fluid, then cavitation can begin.

Vapor pressure

Definition

The vapor pressure of a liquid is the absolute pressure at which the liquid vaporizes or converts into a gas at a specific temperature. Normally, the units are expressed in pounds per square inch absolute (psia). The vapor pressure of a liquid increases with its temperature. For this reason the temperature should be specified for a declared vapor pressure.

At sea level, water normally boils at 212°F. If the pressure should increase above 14.7 psia, as in a boiler or pressure vessel, then the boiling point of the water also increases. If the pressure decreases, then the water's boiling point also decreases. For example in the Andes Mountains at 15,000 ft (4,600 meters) above sea level, normal atmospheric pressure is about 8.3 psia instead of 14.7 psia; water would boil at 184°F.

Inside the pump, the pressure decreases in the eye of the impeller because the fluid velocity increases. For this reason the liquid can boil at a lower pressure. For example, if the absolute pressure at the impeller eye should fall to 1.0 psia, then water could boil or vaporize at about 100°F (see the Tables in Chapter 2 Properties of Water I and II).

Cavitation

Definition

Cavitation is the formation and subsequent collapse or implosion of vapor bubbles in the pump. It occurs because the absolute pressure on the liquid falls below the liquid's vapor pressure.

When the vapor bubbles collapse with enough frequency, it sounds like marbles and rocks are moving through the pump. If the vapor bubbles collapse with enough energy, they can remove metal from the internal casing wall, and leave indent marks appearing like blows from a large ball pein hammer.

This book is dedicated to pumps but we should mention that cavitation could occur in other parts of the pumping system. Under the correct circumstances, valves and pipe elbows are also candidates to suffer damage from cavitation.

The effects of vapor pressure on pump performance

When cavitation occurs in a pump, its efficiency is reduced. It can also cause sudden surges in flow and pressure at the discharge nozzle. The calculation of the NPSHr (the pump's minimum required energy) and the NPSHa (the system's available energy), is based on an understanding of the liquid's absolute vapor pressure.

The effects of cavitation are noise and vibration. If the pump operates under cavitating conditions for enough time, the following can occur:

- Pitting marks on the impeller blades and on the internal volute casing wall of the pump.
- Premature bearing failure.
- Shaft breakage and other fatigue failures in the pump.
- Premature mechanical seal failure.

These problems can be caused by:

- A reduction of pressure at the suction nozzle.

- An increase of the temperature of the pumped liquid.
- An increase in the velocity or flow of the fluid.
- Separation and reduction of the flow due to a change in the viscosity of the liquid.
- Undesirable flow conditions caused by obstructions or sharp elbows in the suction piping.
- The pump is inadequate for the system.

The focus should be on resolving cavitation problems by increasing the external pressure on the fluid or decreasing its vapor pressure. The external pressure could be increased by:

- Increasing the pressure at the pump suction.
- Reducing the energy losses (friction) at the entrance to the pump.
- Using a larger pump.

The vapor pressure of the fluid is decreased by:

- Lowering the temperature of the fluid.
- Changing to a fluid with a lower vapor pressure.

At times, simply removing aspirated air venting the pump will have the same effect.

Cavitation: A practical discussion

Consider the following

I need a pump to raise cold water at 10 gallons per minute. There is an open well with water 40 ft below ground level.

- Do I need a PD Pump?
- Do I need a Centrifugal Pump?
- Should the pump be small, medium, or large?

The reply

No pump in the world can lift cold water 40 ft from an open well in a suction lift condition because the water would evaporate before it comes into the pump. The reason lies in the basic head formula:

$$H = \frac{psi \times 2.31}{sp.\ gr.}$$

$$= \frac{14.7 \text{ psi} \times 2.31}{1.0}$$

$$= 33.9 \text{ ft}$$

You can only raise a column of cold water in a pipe a maximum of 33.9 ft with a pump in suction lift. Beyond 34 ft, the water will boil or vaporize. This is the reason why submersible pumps and vertical turbine pumps exist. There is no limit to the distance you can push a liquid from below, but you can only aspirate a liquid a maximum of 34 ft from below the pump.

AUTHOR'S NOTE

Question: If you put a straw into a glass of milk and suck on the straw, are you really sucking on the milk?

Reply: If you could really suck on the milk, then you wouldn't need the straw. What you're actually doing with your mouth on the straw is lowering the atmospheric pressure inside the straw, so that the atmospheric pressure outside the straw pushes the milk up into your mouth. This is why we say that a pump does not suck. The pump actually generates a zone of low pressure in the eye of the impeller, thereby lowering the atmospheric pressure inside the suction piping. Atmospheric pressure outside the suction piping pushes the liquid up toward the impeller a maximum of 34 ft under ideal circumstances.

Holes in the liquid (cavitation)

A cavitation bubble is a hole in the liquid. If I should have bubbles in the suction of my pump then I have problems. Pumps can move liquid, but they cannot move air or gas bubbles. Compressors exist for moving gases. A gas will not centrifuge. Bubbles occupy space inside the pump and affect the pump's pressure and flow. With vapor bubbles in the low-pressure zones of the pump, the motor's energy is wasted expanding the bubbles instead of bringing more liquid into the pump. As the bubbles pass into the pump's high-pressure zones, the motor's energy is wasted compressing the bubbles instead of expelling the liquid from the pump. The bubbles can collapse as they pass from low- to high-pressure zones in the pump. The water is rather hard.

AUTHOR'S NOTE

You'll know this if you've ever done a belly flop into a swimming pool.

When vapor bubbles collapse inside the pump the liquid strikes the metal parts at the speed of sound. This is the clicking and popping noise we hear from outside the pump when we say that cavitation sounds like pumping marbles and rocks. Sound travels at 4,800 ft per second in water. The velocity head formula gives a close approximation of the energy contained in an imploding cavitation bubble. Remember that implosion is an explosion in the opposite direction.

Using the velocity head formula:

$$H_v = \frac{V^2}{2g}$$

$$= \frac{(4{,}800\ FT/SEC)^2}{2 \times (32.16\ FT)}$$

$$= \frac{23{,}040{,}000}{64.32}$$

$$= 358{,}209\ ft$$

In pump terminology, the approximate energy in an imploding cavitation bubble is 358,209 ft. To convert this energy into pressure:

$$\text{Pressure in psi} = \frac{\text{Head} \times \text{Sp. gr.}}{2.31}$$

$$= \frac{358{,}209 \times 1.0}{2.31}$$

$$= 155{,}069\ psi$$

You can see, based on the velocity head formula, a cavitation bubble impacts the impeller and other pump parts at about 155,069 psi. Other experiments in test laboratories using a more precise rHv, have calculated the impact pressure at 1 Gigapascal, or 147,000 psi. This is the reason that the damage from cavitation appears like someone was beating on your impeller with a large ball pein hammer.

AUTHOR'S NOTE

In medicine, doctors use this same energy contained in cavitation bubbles (Lithotripsy) to treat and destroy kidney stones and tumors. The bubbles act like microscopic jackhammers, disintegrating kidney stones.

If your pump is in cavitation, you'll have one or more of the following:

- Problems with pump packings.
- Problems with mechanical seals.
- Problems with alignment.
- Problems with the bearings.
- Problems with impellers, casings, and wear bands.
- Problems with pump efficiency.
- Problems with leaks and fugitive emissions.

And these problems won't go away until you resolve **cavitation** at its source.

There are five recognized types of cavitation:

- Vaporization cavitation, also called inadequate NPSHa cavitation
- Internal re-circulation cavitation.
- Vane passing syndrome cavitation.
- Air aspiration cavitation.
- Turbulence cavitation.

Let's investigate each of these, their causes and resolutions:

Vaporization cavitation

Vaporization cavitation represents about 70% of all cavitation. Sometimes it's called 'classic cavitation'. At what temperature does water boil? Well, this depends on the pressure. Water will boil if the temperature is high enough. Water will boil if the pressure is low enough.

According to Bernoulli's Law, when velocity goes up, pressure goes down. This was explained in Chapter 1. A centrifugal pump works by acceleration and imparting velocity to the liquid in the eye of the impeller. Under the right conditions, the liquid can boil or vaporize in the eye of the impeller. When this happens we say that the pump is suffering from vaporization cavitation.

This type of cavitation is also called inadequate NPSHa cavitation. To prevent this type of cavitation, the NPSHa in the system (the available energy in the system), must be higher than the NPSHr of the pump (the pump's minimum energy requirement).

A good suggestion to prevent vaporization cavitation is:

NPSHa > NPSHr + 3 ft or more safety margin

AUTHOR'S NOTE

Remember from Chapter 2, the NPSHa formula is: NPSHa = Ha + Hs – Hvp – Hf – Hi. If you want to raise the NPSHa, it will be necessary to increase the elements (Ha, Hs) that add energy to the fluid, or decrease the elements (Hvp, Hf, Hi) that rob energy from the fluid. Also remember that the NPSHr reading, printed on a pump curve, currently represents a point where the pump is already suffering a 3% loss in function due to cavitation. Some people in the industry are calling for a more precise definition of NPSHr, and higher safety margins on NPSHa.

With the pump disassembled in the shop, the damage from vaporization cavitation is seen behind the impeller blades toward the eye of the impeller as illustrated below (Figure 3–1).

To resolve and prevent this type of cavitation damage:

1. Lower the temperature. This reduces the **Hvp**
2. Raise the liquid level in the suction vessel. This elevates the **Hs**.
3. Change the pump.
   - Reduce the speed. This reduces the **Hf**.
   - Increase the diameter of the eye of the impeller. This reduces **Hf** and **Hi**.
   - Use an impeller inducer. This reduces the **Hi**, and increases **Ha**.
   - Use two lower capacity pumps in parallel. This reduces **Hf** and **Hi**.
   - Use a booster pump to feed the principal pump. This increases the **Ha**.

A typical situation often resulting in vaporization cavitation is a boiler

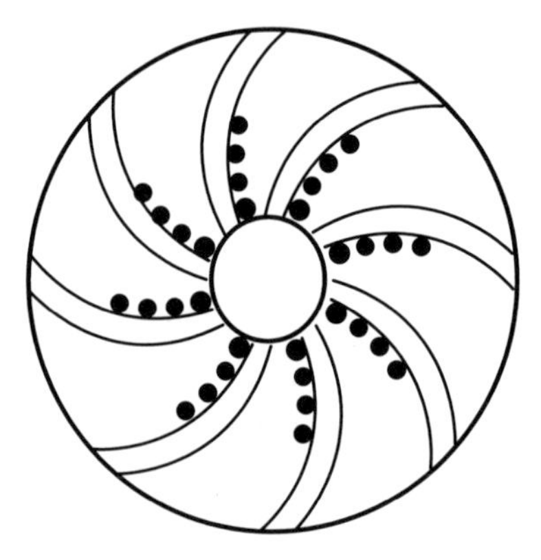

Figure 3–1

feed water pump where the pump drains the deaerator (d-a) tank. Because this pump generates high discharge pressure, it must also generate a strong vacuum in the eye of the impeller. Typically, this pump can generate 200 psi (460 ft of head) and require 30 ft of NPSHr at its duty point.

The d-a tank is normally one of two varieties:

- Vented and exposed to atmosphere.
- Closed and pressurized.

Once again, the formula and elements of NPSHa are:

NPSHa = Ha + Hs – Hvp – Hf – Hi, where:

Ha Atmospheric Head. It is 33.9 ft at sea level.

Hs Static Head. It is the level in the d-a tank above the pump centerline. This is normally about 12 to 15 ft.

Hvp Vapor Head. It is based on the feed water temperature. See Chapter 2, Properties of Water I and II.

Hf Friction Head, or the friction losses in the suction piping. We could assign this a value of 1 ft.

Hi Inlet Head. The losses in the pump suction throat to the impeller eye. These losses could be insignificant up to 2 ft, depending on design.

The feed water in the d-a tank normally runs about 190°F in an open tank. Then we have:

NPSHa = 33.9 + 15 – 22 – 1 – 2
NPSHa = 23.9 ft

if the feed water temperature should be 205°F:

NPSHa = 33.9 + 15 – 30 – 1 – 2
NPSHa = 15.9 ft

if the d-a tank should be closed, then the Ha = Hvp. Therefore they cancel:

NPSHa = 15 – 1 – 2
NPSHa = 12 ft

If the d-a tank should be sealed and artificially pressurized with steam gas (sometimes they are, or should be), then each 10 psi adds 23.1 ft of artificial head to the system's NPSHa. If the boiler feed water pump has a NPSHr of 30 ft at the duty point, now you can see why boiler feed water pumps are sometimes considered problematic regarding cavitation. This also demonstrates the need to seal and artificially pressurize d-a tanks to get the NPSHa above the NPSHr of the pump.

Internal re-circulation

This is a low flow condition where the discharge flow of the pump is restricted and the product cannot leave the pump. The liquid is forced to re-circulate from high-pressure zones in the pump into low-pressure zones across the impeller.

This type of cavitation originates from two sources. First, the liquid is circulating inside the volute of the pump at the speed of the motor and it rapidly overheats. Second, the liquid is forced to pass through tight tolerances at very high speed. (These tight tolerances are across the wear bands on enclosed impellers, and between the impeller's leading edges and the volute casing on opened impellers.) The heat and the high velocity cause the liquid to vaporize.

With the pump disassembled in the shop, with open impellers, the damage is seen on the leading edge of the impeller blades toward the eye of the impeller, and on the blade tips toward the impeller's OD. With enclosed impellers, the damage reveals itself on the wear bands between the impeller and the volute casing. See the illustration (Figure 3–2).

To correct this condition with an opened impeller, it's necessary to perform an impeller adjustment to correct the strict tolerance between the blades and the volute. Some back pullout pumps are designed with jack bolts on the power end of the bearing housing to easily perform this adjustment without pump disassembly.

This condition cannot be corrected on pumps with an enclosed impeller. You need to relax the restricted discharge flow on the pump. The problem could be a clogged downstream filter, a closed discharge valve, an over-pressurized header (back-pressurizing the pump), or a

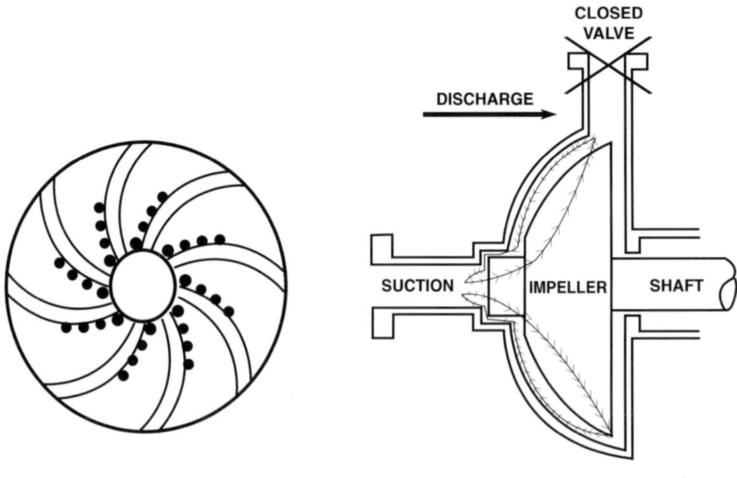

Figure 3–2

check valve installed backwards, or operating the pump at or close to shut-off head. Maybe a variable speed motor could help under certain circumstances.

The vane passing syndrome

The vane passing syndrome can exist when the blade tips at the OD of the impeller are passing too close to the cutwater on the pump casing. This can be caused by exchanging an impeller for a larger diameter impeller, or from re-metalizing or coating the internal housing of the pump. The free space between the impeller blade tips and the cutwater should be 4% of the impeller diameter (Figure 3–3).

With the pump disassembled on the shop table, the damage is seen on the blade tips at the OD of the impeller, and just behind the cutwater on the internal volute wall.

For a 13″ impeller, the free space should be 4% of the impeller diameter between the blade tips and the cutwater.

$$13'' \times 0.04 = .520''$$

On a 13″ impeller, there should be at least a half-inch free space between the blade tips as they spin past the pump cutwater.

Air aspiration

Air can be drawn into the piping and pump from diverse forms and different points. Air can enter into the piping when the pump is in vacuum. An example of this is a lift pump. Lift pumps tend to lose their prime and aspirate air into the suction piping and pump.

1. The air can come into the pump through:
   - The pump shaft packing.
   - Valve stem packings on valves in the suction piping.

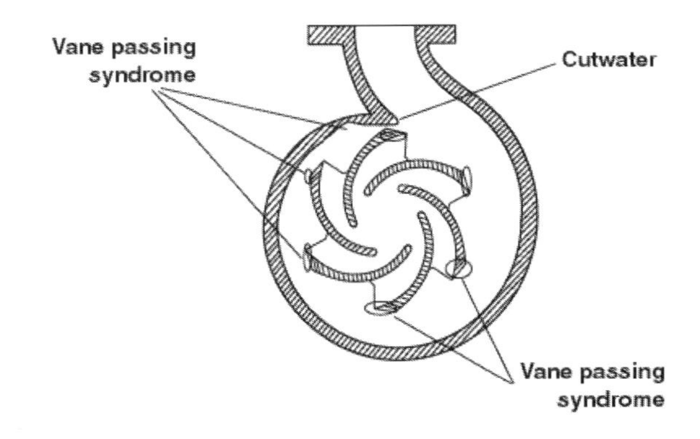

Figure 3–3

- Joint rings on suction piping.
- Flange face sheet gaskets at pipe joints.
- O-rings and threaded fittings on instrumentation in the suction piping.
- O-rings and other secondary seals on single mechanical seals.
- The faces of single mechanical seals.

2. Air can also enter into the pump from bubbles and air pockets in the suction piping.
3. Products that foam can introduce air into the pump.

The evidence of this type of cavitation manifests itself like vaporization cavitation, or inadequate NPSHa. The damage to the impeller appears like vaporization cavitation. **However, the solution is different**.

To prevent this type of cavitation, you need to seal all points of entrance and escape.

1. Tighten all flange faces and gaskets.
2. Tighten all pump packing rings and all valve stem packings on suction piping.
3. Keep the velocity of the fluid in the suction piping at less than 8 ft per second. It may be necessary to increase the diameter of the pipe.
4. Consider using dual mechanical seals with a forced circulation barrier fluid (not induced with a pumping ring) between both sets of faces on:
   - Vertical pumps.
   - Lift pumps and pumps in suction lift conditions.
   - Pumps in vacuum.
   - Pumps operating to the right of their best efficiency point (BEP). This is explained in Chapters 7 and 8.

Turbulence cavitation

This is cavitation due to turbulence caused by the following:

1. Formation of vortexes in the suction flow.
2. Inadequate piping, sharp elbows, restrictions, connections, filters and strainers in the suction.
3. The waterfall effect in suction vessels.
4. Violating or not respecting the submergence laws.

With the pump disassembled on a shop table, the evidence of turbulence cavitation appears like vaporization cavitation or inadequate NPSHa. To deal with these problems, the technician must understand the **'Lost Art of Pipefitting'**. There is a complete discussion on the laws of submergence and turbulence in Chapter 17 of this book.

Review for preventing cavitation

The general rule is

$$NPSHa > NPSHr + 3\ ft$$

To increase the NPSHa

1. Raise the level in the suction vessel. This increases Hs.
2. Elevate the suction vessel. This increases Hs.
3. Lower the pump. This increases Hs.
4. Reduce the friction in the suction piping. This is probably the most creative way to deal with cavitation. This reduces the **Hf**.
   - Use larger diameter suction pipe.
   - Change the pipe schedule. If there is a designated schedule, you can bet it was based on discharge pressures and not suction pressures.
   - Change to pipe with lower friction characteristics. Ex. Change cast iron piping for PVC or even food grade stainless.
   - Move the pump closer to the suction vessel.
   - Convert globe valves into gate valves if possible.
   - Convert quarter turn butterfly valves into ball valves.
   - Be sure all ball valves are full port design.
   - Be sure all suction valves are totally open.
   - Reduce multiple elbows. If a system is designed with 9 or more elbows in the suction piping, you can be sure that some of these elbows are self-canceling. If so, then some elbows can be eliminated.
   - Convert 2 or 3 close fitting elbows into a flexible 'S'.
   - Convert 'mitered 90° elbows' and short radius elbows into long radius elbows.
   - Keep suction pipe inside diameters clean and scale free.

- Change filters and strainers with more frequency.
- Be sure all pipe gaskets and ring seals are perfectly centered within the flange faces.

5. Lower the temperature of the fluid in the suction vessel. This decreases the Hvp.
6. Pressurize the suction vessel. This increases the artificial **Ha** 23 ft for every 10 psi.

To Reduce the pump's NPSHr

1. Use a pump with a larger suction flange. This lowers the **Hi**. An example of this would be to change a 3 × 4 × 10 pump into a 4 × 6 × 10 pump. The 10-inch impeller needs to remain the same for discharge pressure. However, by converting the 4-inch suction flange into a 6-inch suction flange, the inlet losses would be reduced.
2. Machine and polish the suction throat of the pump. This is probably the worst casting, and roughest finish in the entire pump. Center the suction flange on a lathe and ream-out the suction throat. This reduces the **Hi**.
3. Machine open and increase the inside diameter of the eye of enclosed impellers. This reduces the **Hi**.
4. Use a larger/slower pump. This reduces the **Hi** and **Hf**.
5. Use a small booster pump to feed the principle pump. This increases the artificial head (**Ha**).
6. Use smaller capacity pumps in parallel. This reduces the **Hi** and **Hf**.
7. Use a double suction impeller. Convert an end suction centrifugal pump into a split case horizontal design.
8. Use an impeller inducer.

As you can see by reading through some of these solutions to cavitation, some of the changes are very practical, and others are not.

AUTHOR'S NOTE

A few of the above-mentioned solutions to cavitation are almost comical and not even cost effective. The idea is that they would work to reduce and stop cavitation and the resulting seal, bearing and pump failure. Too many maintenance people (engineers and mechanics) are running around in circles, wringing their hands, and jumping up and down, trying to deal with cavitation. Who would have thought that there are so many solutions, practical or not?

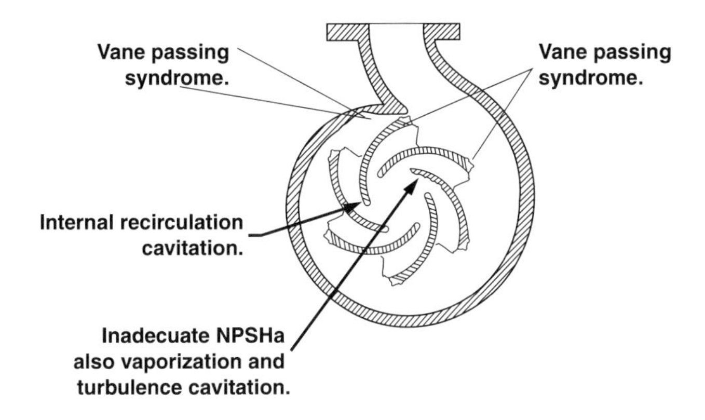

Figure 3–4

Figure 3–4 shows the inside of the pump where cavitation occurs.

Cavitation review

Reviewing the cavitation theme, we observe that the majority of cavitation is caused or induced by operation and design.

As we've discussed, system design is responsible for much of cavitation. Yet, the maintenance mechanic is responsible for stopping and preventing cavitation. And certainly, it's the maintenance mechanic who has to deal with the results of cavitation, the constant changing of bearings, mechanical seals, damaged impellers, wear rings and other pump parts.

Most mechanics have seen the damaged pump parts with the pump torn apart on the shop table. Almost all mechanics have seen the scratches, gouges, and track marks of cavitation on the impeller blades and pump housings. But without understanding and analyzing the causes of cavitation damage, the mechanic can only install more bearings and seals, and possibly hide the damage by re-metalizing the housing and impeller. This is somewhat frustrating for the mechanic, because he or she will be performing the same repair work in the next few weeks or months.

Do something about cavitation!!

The worst thing you can do is....nothing. Do something. Take this information, go back into your process plant, and you can prevent and stop the majority of cavitation.

Type	Cause	Explanation
Vaporization	Operation / Design	Vaporization (inadequate NPSHa) is the result of something that happens in the suction side of the pump, strangled valve, clogged filter, inadequate submergence.
Internal Re-circulation	Operation / Design / Maintenance	Internal Re-circulation results from something in the discharge side of the pump: strangled valve, clogged filter, over-pressurized header, check-valve installed backwards.
Vane Pass Syndrome	Design / Maintenance	Oversized impeller, inadequate aftermarket parts, repair or rebuild with incorrect specifications and measurements.
Air Aspiration	Design / Operation / Maintenance	Design, Improper suction pipe, fluid flow too fast. Inadequate flange torque, operation to the right of BEP.
Turbulence	Design	Design, Inappropriate suction system (vessel, piping, connections, fittings) design.

Be aware that in some cases, you'll have to live with cavitation. Many pumps suffer cavitation for reasons of inadequate design. For example, when operating only one pump in a parallel system, this pump tends to go into cavitation. Pumps that perform more than one duty through a valve manifold tend to suffer cavitation. Pumps that fill and drain tanks from the bottom tend to suffer cavitation. The last pump drawing on a suction header tends to cavitate. And of course vacuum pumps and pumps in a high suction lift are candidates for cavitation.

Some solutions may not be practical, or economical, or timely and consistent with production. You could be forced to live with cavitation until the next plant shutdown to make the necessary corrections. In the meantime, the cavitation shock waves and vibrations will travel through the impeller, down the shaft to the mechanical seal faces, and onto the shaft bearings. We offer some specific recommendations for surviving cavitation shock waves and vibrations in Chapters 13 and 14 on Mechanical Seals.

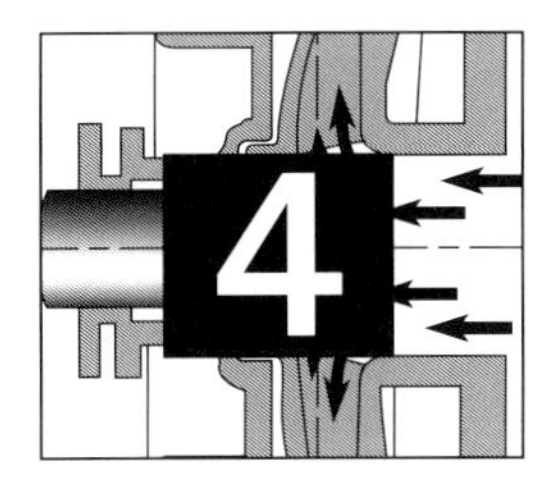

The Affinity Laws

Introduction

The Affinity Laws are a group of rules that govern your pumps. Sometimes they're called the Laws of Similarity. In years gone by, these laws served to predict a pump's operation when exported to a country with different electricity. The electricity in the United States is 60 hertz (Hz). This means that the electricity moves in waves with a frequency of 60 cycles per second. Some other countries in the world have electricity moving at a frequency of 50 Hz., or 50 cycles per second. A change in frequency brings about a change in running speed (rpm). Upon exporting a pump into a country and connecting it to an electric motor running at different speeds, the pump has different operational characteristics. Those characteristics change by the Affinity Laws. Now, with the arrival of the variable speed electric motor, called Variable Frequency Drive (VFD), the Affinity Laws will become increasingly important with industrial pumps.

The Laws

When a pump's velocity changes, measured in revolutions per minute (rpm), the operational characteristics also change. These changes can be calculated using the Affinity Laws. Before continuing, let's define some terms we'll be using:

1. Flow = Capacity = Q: Liquid volume measured in gallons per minute (gpm) or liters per minute, cubic meters per hour, or some other rate.
2. Head = H: Liquid force measured in feet of elevation. H can be converted into pounds per square inch (psi). This is discussed in Chapter 2.

3. Brake Horsepower = BHP: Energy needed to pump a liquid.
4. Speed = Velocity = N: Shaft speed measured in revolutions per minute (rpm).

Stated simply, the Affinity Laws indicate:

- Flow, Q, changes directly proportional to a change in velocity. $Q \alpha N$.
- Head, H, changes directly proportional with the **square** of the change in velocity, $H \alpha N^2$.
- Power, BHP, changes directly proportional with the **cube** of the change in velocity, $BHP \alpha N^3$.
- α means 'directly proportional to'.

In the form of equations:

$$\text{New Flow} = \text{Initial Flow} \times \left(\frac{\text{r.p.m. New}}{\text{r.p.m. initial}}\right) \qquad \frac{Q1}{Q2} = \frac{N1}{N2}$$

$$\text{New Head} = \text{Initial Head} \times \left(\frac{\text{r.p.m. New}}{\text{r.p.m. initial}}\right)^2 \qquad \frac{H1}{N2} = \left(\frac{H1}{N2}\right)^2$$

$$\text{New BHp} = \text{Initial BHp} \times \left(\frac{\text{r.p.m. New}}{\text{r.p.m. initial}}\right)^3 \qquad \frac{BHp1}{BHp2} = \left(\frac{N1}{N2}\right)^3$$

With a small change in velocity (say 20 or 50 rpm), the pump's efficiency would not be affected. But with a velocity change of say twice the speed, or one half the speed, you could expect a small change in the pump's efficiency of maybe 2 or 3%. The pump's efficiency would tend to increase by a couple of points at twice the speed, and the efficiency would tend to decrease a couple of points at one half the speed. In a practical situation, a process engineer or operator might double the speed of the electric motor because the plant is undergoing an expansion and needs twice the production. That engineer or operator should know that:

2 × velocity = 2 × flow

= 2 × capacity

= 2 × gpm

= 2 × production

2 × velocity = 2^2 = 4 × Head
= 2^2 = 4 × Pressure (psi)
= 2^2 = 4 × NPSHr
= 2^2 = 4 × Misalignment in the bearings.

2 × velocity = 2^3 = 8 × (BHp) requirements
= 2^3 = 8 × maintenance costs
= 2^3 = 8 × downtime
= 2^3 = 8 × erosion in pipes and elbows
= 2^3 = 8 × impeller wear
= 2^3 = 8 × wear in wear rings
= 2^3 = 8 × other close tolerance wear
= 2^3 = 8 × friction losses (Hf) in pipes
= 2^3 = 8 × Hf in fittings, valves, etc.

Consequently, if the pump and motor speed were reduced by 50%:

- The flow would be divided by 2.
- The head would be divided by 4 (2^2×)
- The BHP would be divided by 8 (2^3×)

The relationship between velocity and horsepower requirement (BHp) presents some good arguments in favor of variable speed motors, VFDs. When normal manufacturing plant operations don't depend on time (for example, if you have all night to drain a tank) the pump operator can perform this function at 50% speed, while consuming one-eighth the BHp. VFDs are practical in manufacturing plants because they permit the pumps and other equipment to work at their best efficiency. Controlling pump flow with a VFD is better than using a constant speed motor and controlling the flow by opening and closing valves.

VFDs work well on most PD pumps, and also centrifugal pumps performing a flow service. Remember that some centrifugal pumps are required to comply with head, pressure and elevation applications. An example of this would be a boiler feed water pump, or a pump pushing a fluid through a filter. In these applications, the VFD may only be effective at 85% to 100% maximum speed. Incorrect use of the VFD or lack of understanding of the affinity laws could prejudice these applications because running the VFD and pump at 50% speed would only generate one-fourth the head, pressure, or elevation.

The Affinity Laws and the impeller diameter

If the velocity should remain fixed, the flow, head and BHP will change when the impeller diameter changes. With a change in the impeller diameter (this is called an impeller 'trim'), the affinity laws indicate:

- The Flow changes directly proportional to the change in diameter, $Q \alpha D$.
- The Head changes directly proportional with the square of the change in the impeller diameter, $H \alpha D^2$.
- The BHp changes directly proportional with the cube of the change in the impeller diameter, $BHp \alpha D^3$.

In the form of Equations:

$$\text{New Flow} = \text{Initial Flow} \times \left(\frac{\text{New Diameter}}{\text{Initial Diameter}}\right) \qquad \frac{Q1}{Q2} = \frac{D1}{D2}$$

$$\text{New Head} = \text{Initial Head} \times \left(\frac{\text{New Diameter}}{\text{Initial Diameter}}\right)^2 \qquad \frac{H1}{H2} = \left(\frac{D1}{D2}\right)^2$$

$$\text{New BHp} = \text{Initial BHp} \times \left(\frac{\text{New Diameter}}{\text{Initial Diameter}}\right)^3 \qquad \frac{BHp1}{BHp2} = \left(\frac{D1}{D2}\right)^3$$

What's the practical application of these laws?

The pump curve, the H-Q curve, is in a descending profile. This means that with an increase in flow 'Q', gpm, the Head 'H', or pressure falls. And if flow is reduced, the pressure rises. At times, in normal industrial production, the flow must rise and fall, but the pressure or head must remain a constant.

Many industrial processes experience seasonal rises and falls. This means that flow varies. For example, we consume more cough syrup and aspirin in the flu season and less of these products in other seasons.

(This chapter is written with the authors suffering this ailment.) Normally we buy more ice cream in summer, and less in winter.

The pasteurization process for milk and ice cream requires heating the milk to a specified temperature and pressure for a specified time to kill all germs and bacteria in the milk. This pressure is constant although the production of milk and ice cream goes up and down with consumption and the seasonal changes. It is the same with the treatment of potable water, sterile pharmaceuticals, and petroleum refining.

Let's consider sterile water, used in the preparation of medications for injection. A typical process to sterilize water would require boiling the water at 35 psi, and pumping the water at 40 gpm to 70 gpm, according to consumption. The 35 psi is a constant for the water to pass through the heat exchanger, and a bank of filters. To compensate for the change in demand for sterile water, the affinity laws are used, varying the diameter of the impeller, so that the pump can pump 40 gpm at 35 psi, or 50 gpm at 35 psi, or 70 gpm at 35 psi. This allows the operator to use the same pump and motor, and only change the impeller diameter depending on the needs of production. This precise manipulation of pumping parameters could not be obtained by opening and closing valves, or by simply controlling the pump speed with a VFD.

AUTHOR'S NOTE

Most people change their wardrobe and clothes as the weather changes. Most people would change their cars if their transportation needs change. As the authors of this book, it has always seemed strange to us that most pumps are sold with only one impeller. There is absolutely nothing wrong in selling (or buying) a pump with various impellers of different diameters, ready to be changed when the needs of production change seasonally, or with an advertising campaign. This is the reason that back pullout pumps exist.

Many clients specify and buy pumps with the back pullout option, and they never take advantage of the option. This is like buying a car with an air conditioner and never turning it on. Many engineers, operators, and even pump salesmen believe that the back pullout feature is designed to facilitate maintenance. This is wrong. The back pullout pump exists to facilitate the rapid and frequent impeller change, adapting the pump to the ever-changing needs of production. The back pullout pump exists to facilitate production.

Manipulating flow and controlling pressure by varying the impeller diameter conserves kilowatts of energy, and this is the third affinity law in this group. A pump consuming 10 BHP with a 10 inch impeller, would only consume 7.3 horses with a 9 inch impeller.

This means a 10% reduction in the impeller diameter, would bring about almost 30% reduction in energy. These energy savings will easily cover the cost of multiple impellers and the manpower to change them frequently.

Useful Work and Pump Efficiency

Useful work from a pump

The physicist James Watt is honored in the electrical community for the term 'watt'. He made various advancements and improvements to stationary boilers and steam engines. It is said that the first practical use of the steam engine was in raising (call it pumping) water out of the coalmines. Almost all mines would flood if the water were not pumped from the bilge, out of the mine. Before the steam engine, the miners used children and horses to lift and carry the bilge water.

James Watt developed the terms of energy, work, and power. He defined the following:

- Energy is the capacity to perform work. Example; I have the energy in my bicep muscle to lift a 100-pound weight.
- Work is a force multiplied over a distance. Example: If I lift a 5-pound weight one foot into the air, then I've performed 5 foot-pounds of work.
- Power is work performed within a certain specified time frame. Power is when I perform 5 foot-pounds of work within a second, or minute.

Many people confuse these terms, but they actually have precise definitions. If I should lift 10 pounds a distance of 10 feet, then I've performed 100-foot-pounds of work (10 pounds × 10 feet = 100). Before the steam engine, the most powerful force to perform work, or exert a force, was a horse.

James Watt, with actual tests, determined that a coal mine draft horse could lift 550 pounds, a distance of one foot, within a second. So, James Watt declared 550 foot-lbs/sec. to be one Horsepower. To this day, this has become the standard definition of a horsepower (1 HP =

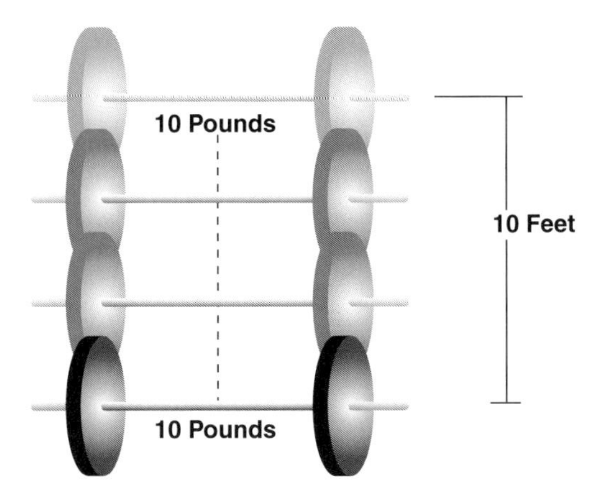

Figure 5-1

550 ft.-lbs./sec.). This is the reason that even today, all motors, whether steam, internal combustion engines, boilers, electric motors, gas turbines, and even jet and rocket engines are rated in Horsepower, and not Ostrich power or Iguana power.

We say that the motor generates horsepower (HP), and that the pump consumes brake horsepower (BHp). The difference between HP (output) and BHp (input) is what is lost in the power transmission; the bearings, shaft, and coupling between the motor and the pump.

We say that the useful work of the pump is called Water horsepower (WHp). It is demonstrated mathematically as:

$$\mathbf{WHp} = \frac{\mathbf{H \times Q \times sp.gr.}}{\mathbf{3960}}$$

Where: **H** = head in feet generated by the pump **Q** = flow recorded in gallons per minute **sp. gr.** = specific gravity
3960 = constant to convert BHp into gallons per minute

$$\mathbf{3960} = \frac{\textbf{Horsepower} \times \textbf{60 secs. / min.}}{\textbf{Weight of 1 gal. of water}}$$

$$\mathbf{3960} = \frac{\textbf{550 lbs. ft / secs.} \times \textbf{60 secs.}}{\textbf{8.333 lbs. / gal.}}$$

If the pump were 100% efficient, then the BHp would be equal to the WHp. However, the pump is not 100% efficient so the BHp = WHp × efficiency, and the formula is:

$$\mathbf{BHp} = \frac{\mathbf{H \times Q \times sp.gr.}}{\mathbf{3960 \times eff.}}$$

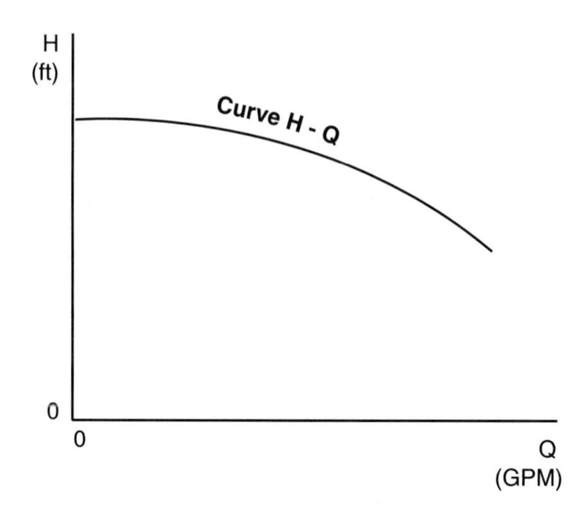

Figure 5–2

The graph (Figure 5–2) shows the useful work of a pump. Notice that the pump pumps a combination of head and flow. As a general rule, as flow increases, the head decreases.

Example:

Given: Pressure or Head required = 100 feet at 200 gpm. What is the water horsepower required for this pump? Assume a sp. gr. of 1.0

$$\mathbf{WHp} = \frac{\mathbf{H \times Q}}{\mathbf{3960}} = \frac{\mathbf{100\ ft. \times 200\ gpm}}{\mathbf{3960}} = \mathbf{5.05\ HP}$$

If the specific gravity at pumping temperature were not equal to 1.0, then the water horsepower would be adjusted by the specific gravity.

$$\mathbf{WHp} = \frac{\mathbf{H \times Q \times sp.gr.}}{\mathbf{3960}}$$

Flow determination

Flow is the number of gallons per minute that the pump will discharge.

- Any pump will generate more flow as the discharge pressure is reduced.
- Equally, the pump will generate less flow as the discharge head or pressure requirements are increased. Obviously, both flow and head should be known before selecting a centrifugal pump.

AUTHOR'S NOTE

It is not practical to declare the flow without the accompanying head requirements. For this reason, when someone asks for the pump specifications, they need to know the flow in gallons per minute and the head in feet.

- The available areas in the impeller, and the available area in the volute determine the flow, gpm. There are two critical areas in the impeller, the exit area and the entrance area. For the volute casing, the most important area is the 'cutwater'. All fluid must pass this point.
- Head or pressure is developed in the pump; when the impeller imparts rotational energy to the liquid (increasing the liquid's velocity), and then the volute converts this energy (by decreasing the velocity) into pressure.
- The relationship between the 'exit area' of the impeller, and the 'cutwater area' of the volute, generally determine the flow of the pump.

See the illustration below (Figure 5–3):

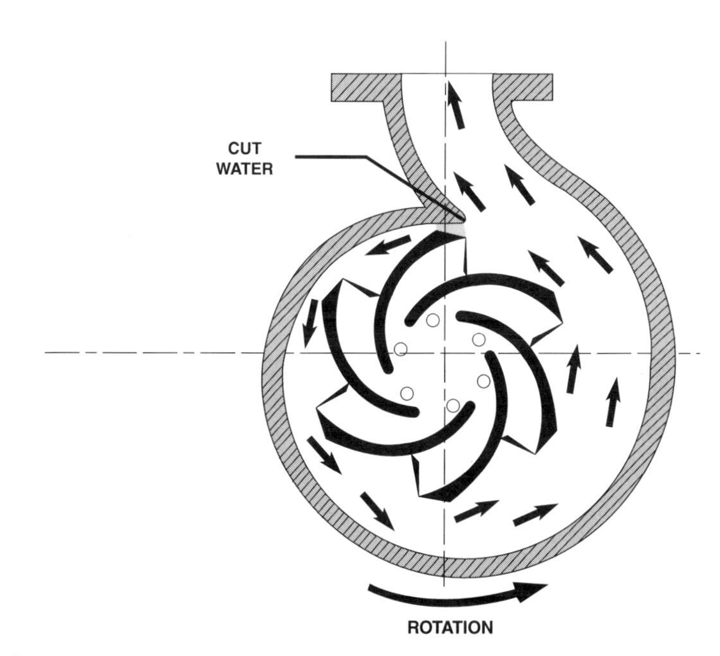

Figure 5–3

Pump efficiency

Numerous factors affect the pump's efficiency. The impeller is one of the most important efficiency factors.

Affecting the impeller's behavior are:

1. The impeller velocity.
2. The impeller diameter.
3. The number of blades on the impeller.
4. The diameter of the eye of the impeller.
5. The thickness of the impeller.
6. The pitch (angle) of the blades.

Factors that affect the efficiency

1. Surface finish of internal surfaces – Efficiency increases from better surface finishes are mostly attributable to the specific speed Ns (discussed in Chapter 6) of the pump. Generally, the improvements in surface finishes are economically justifiable in pumps with low specific speeds.
2. Wear ring tolerance – Close tolerances on the wear rings have a tremendous effect on the pump's efficiency, particularly for pumps with a low specific speed (Ns < 1500).
3. Mechanical losses – Bearings, lip seals, mechanical seals, packings, etc., all consume energy and reduce the pump's efficiency. Small pumps (less than 15 HP) are particularly susceptible.
4. Impeller diameter – There will be an efficiency reduction with a reduction in the impeller diameter. For this reason, it's not recommended to reduce (trim) the impeller by more than 20%. For example, if a pump takes a full sized 10-inch impeller, don't trim the impeller to less than 8-inches diameter. This would be a 20% reduction.
5. Viscosity – Viscous liquids generally have a prejudicial effect on efficiency. As the viscosity of the fluid goes up, generally the efficiency of most pumps goes down. There are exceptions.
6. Size of solid particles – Low solids concentrations (less than 10% average) classified by size and material, generally exhibit no adverse affect to pump efficiency. However, the discharge configuration of the pump must be sufficiently large to prevent obstructions. For example, sanitary and wastewater pumps that handle high solids,

have 2 or 3 blades on a specially designed impeller with lower efficiency.

7. The type of pump – There are many types of pumps with configurations and characteristics for special services, such as sanitary, wastewater, and solids handling, etc., taking into account the Ns and design that perform their services effectively with a slightly less than optimum efficiency. In simple terms, special designs and services generally reduce efficiencies.

$$\textbf{Efficiency} = \frac{\textbf{Work Output}}{\textbf{Work Input}} = \textbf{Power Produced}$$

$$\textbf{Pump Efficiency} = \frac{\textbf{Water Horsepower}}{\textbf{Brake Horsepower}} = \frac{\textbf{WHp}}{\textbf{BHp}}$$

$$\textbf{Pump Efficiency} = \frac{H \times Q \times sp.\ gr.}{3960 \times BHp}$$

$$\textbf{Coupling Efficiency} = \frac{\textbf{Pump Horsepower}}{\textbf{Motor Horsepower}} = \frac{\textbf{BHp}}{\textbf{Hp}}$$

$$\textbf{Motor Efficiency} = \frac{\textbf{Motor Horsepower Output}}{\textbf{Energy / Power Input}} = \frac{\textbf{Hp}}{\textbf{KW}}$$

$$BHp = \frac{H \times Q \times sp.\ gr.}{3960 \times eff.}$$

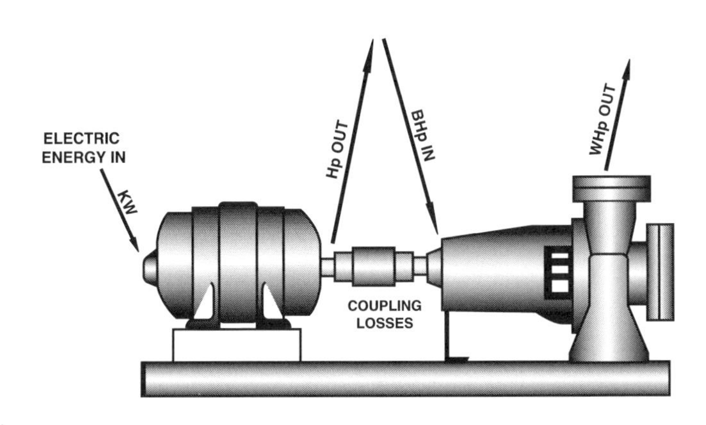

Figure 5-4

Calculating pump efficiency

Example

A system requires 2,500 gpm flow of brine (salt water with sp. gr. of 1.07) at 120 psi., 213 BHp required.

Calculate Head

$$\text{Head} = \frac{\text{psi} \times 2.31}{\text{sp. gr.}} = \frac{120 \text{ psi} \times 2.31}{1.07} = 259.06 \text{ Feet}$$

Calculate Efficiency:

$$\text{Efficiency} = \frac{H \times Q \times \text{sp. gr.}}{3960 \times \text{BHp.}} = \frac{259 \text{ ft.} \times 2500 \text{ gpm} \times 1.07 \text{ sp. gr.}}{3960 \times 213 \text{ BHp}} = 82\%$$

This pump is 82% Efficient.

Pump Classification

Introduction

In Figure 6–1, Pump Classification, we see two principal families of pumps: Kinetic Energy pumps and Positive Displacement pumps. These two families are further divided into smaller groups for specific services. Both pump families complete the same function, that is to add energy to the liquid, moving it through a pipeline and increasing the pressure, but they do it differently.

Positive displacement pumps

Positive Displacement pumps perform work by expanding and then compressing a cavity, space, or moveable boundary within the pump. In most cases, these pumps actually capture the liquid and physically transport it through the pump to the discharge nozzle. Inside the pump where the cavity expands, a zone of low pressure, or vacuum, is generated that causes the liquid to enter through the suction nozzle. Then the pump captures and transports the liquid toward the discharge nozzle where the expanded cavity compresses. In this sense, because the available volume of space at any point inside the pump is a constant, we can say that in theory, these pumps are considered a 'constant volume device' with every revolution or reciprocating cycle.

Theoretically, the curve of a Positive Displacement pump should appear as in (Figure 6–2).

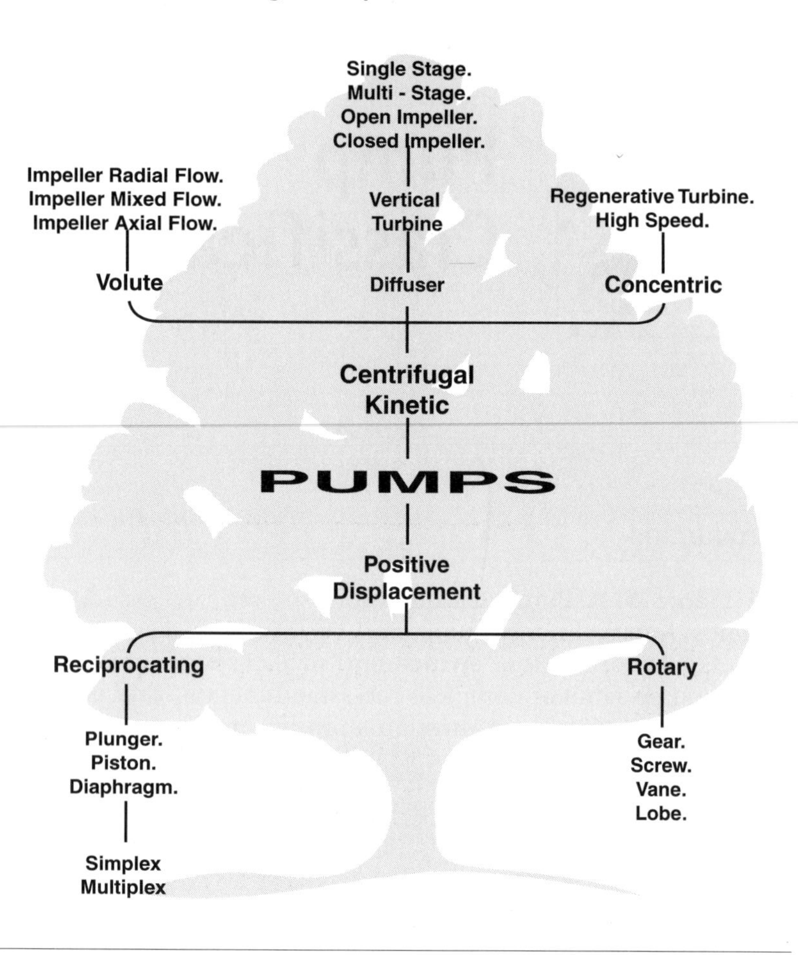

Figure 6–1

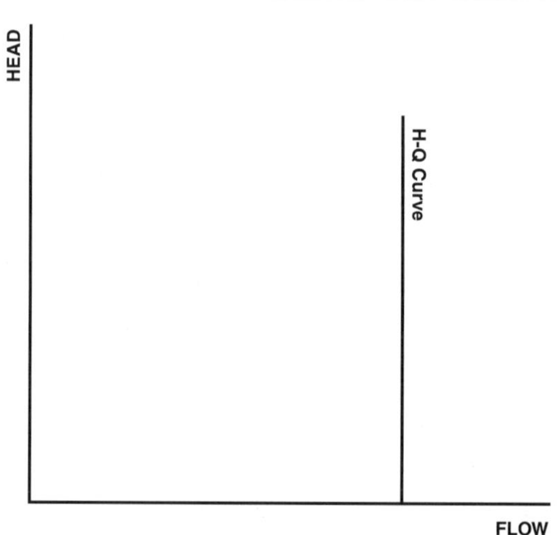

Figure 6–2

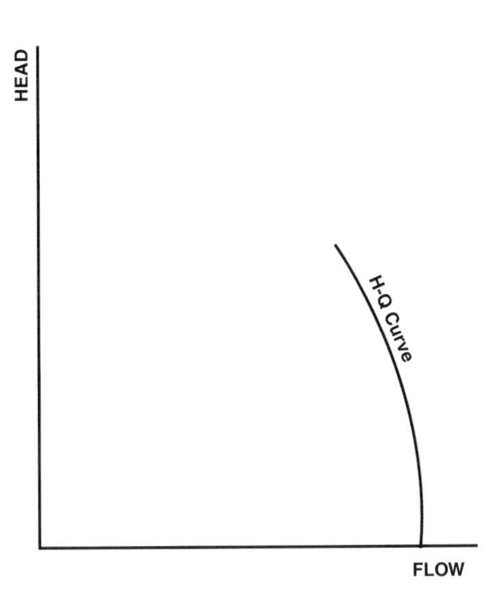

Figure 6–3

In reality, there are small losses in volume delivered as the pressure or resistance increases, so a more representative PD pump curve appears in Figure 6–3.

The flow through a PD pump is mostly a function of the speed of the driver or motor. It is important to note that a pump cannot generate flow. The flow must be available to the pump suction nozzle. In this sense the flow in a PD pump is actually energy, called net positive inlet pressure. The pressure or head that a PD pump can generate is mostly a function of the thickness of the casing and the strength of the associated accompanying parts (seals, hoses, gaskets).

Positive displacement pumps normally have some strict tolerance parts. These parts vary with the type and design of the pump. This strict tolerance controls the flow, and the pressure that these pumps can generate. When this tolerance opens or wears by just a few ten thousandths, these pumps lose almost all their efficiency and ability to function. These strict tolerance parts must be changed with a planned certain frequency, based on the abrasive nature and lubricity of the pumped fluid, to maintain the maximum efficiency of the pump.

There is no definite demarcation line, but positive displacement pumps normally are preferred over centrifugal pumps in applications of:

- Viscous liquids,
- Precise metering, (dosification, pharmaceutical chemistry) and
- Where pressures are high with little flow.

Centrifugal pumps

Centrifugal pumps perform the same function as PD pumps, but they do it differently. These pumps generate pressure by accelerating, and then decelerating the movement of the fluid through the pump.

The flow, or gallons per minute, must be available to the pump's suction nozzle. This flow, or energy, in centrifugal pumps is called NPSH or Net Positive Suction Head (discussed in Chapter 2). These pumps, like their PD sisters, cannot generate flow. No pump in the world can turn three gallons per minute at the suction nozzle, into four gallons per minute out of the discharge nozzle. The fluid enters into through the suction nozzle of the pump to the eye of the impeller. The fluid is trapped between the veins or blades of the impeller. The impeller is spinning at the velocity of the driver. As the fluid passes from the eye, through the blades toward the outside diameter of the impeller, the fluid undergoes a rapid and explosive increase in velocity. Bernoulli's Law states that as velocity goes up, the pressure goes down, and indeed there is a low-pressure zone in the eye of the impeller. The liquid that leaves the outer diameter of the impeller immediately slams into the internal casing wall of the volute, where it comes to an abrupt halt while it collects in the ever-expanding exit chamber of the volute. By Bernoulli's law, as velocity goes down, the pressure increases. The velocity is now converted into head or pressure available at the discharge nozzle. Because the impeller diameter and motor speed is mostly constant, the centrifugal pump can be considered to be a constant head or pressure device. The theoretical curve of the centrifugal pump is seen in Figure 6–4.

In reality, these pumps lose some head (pressure) as energy is channeled

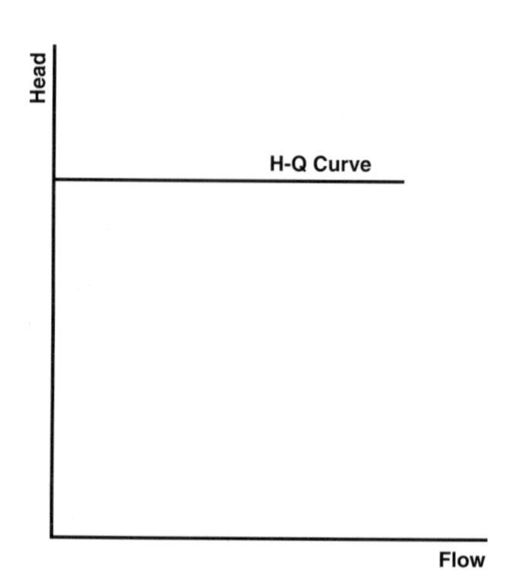

Figure 6–4

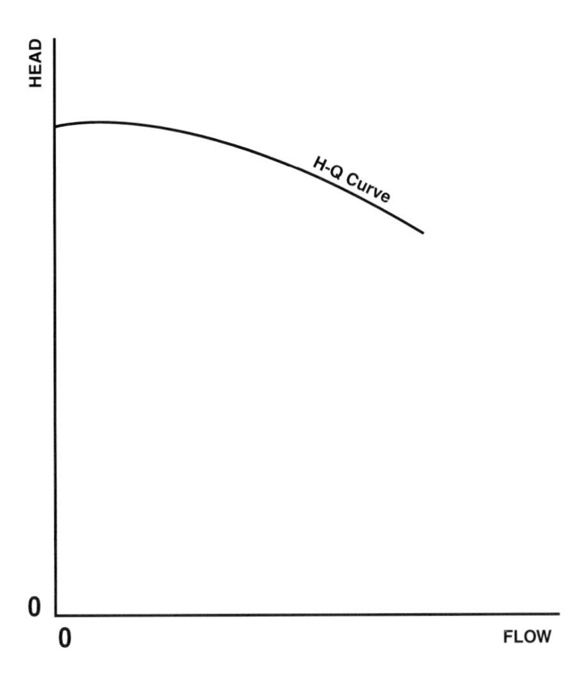

Figure 6–5

toward increasing the flow and speed. A more realistic curve would appear as in Figure 6–5.

Conceptual difference

This is the conceptual difference between centrifugal pumps and positive displacement pumps, as you can see when we superimpose the four theoretical and realistic curves (Figure 6–6).

Upon developing the system curve, the pump curve will always intersect the system curve, it doesn't matter about the pump design. We'll see this later in Chapter 8.

According to Figure 6–1, entitled Pump Classification, approximately half of all existing pumps are centrifugal, and the other half are PD-pumps. Actually, it's possible that there are more PD pump designs than centrifugal pump designs, and a higher population of PD pumps in the world for specific applications, than centrifugal pumps in general applications.

However, in heavy industry, meaning metallurgical (steel and aluminum) processes, mining, petroleum refining, pulp and paper production, and the process industries like chemical and pharmaceutical production, potable water, wastewater, edible products, and manu-

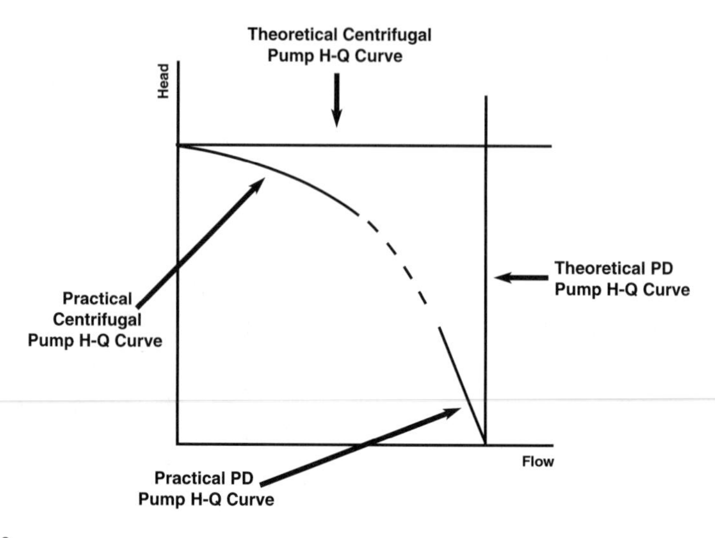

Figure 6–6

facturing in general, we observe that there are more centrifugal pumps. About 90% of the pumps in industry are centrifugal pumps. And the PD pumps found in industry have much in common with their centrifugal sisters. They are mostly rotary designed pumps, with precision bearings and a shaft seal. Much of the theory, and system needs, are applicable to both types of pumps.

Centrifugal volute pumps

This type of pump adds pressure to a liquid by manipulating its velocity with centrifugal force, and then transforms the force into pressure through the volute. By observing Figure 6–7, we see that the liquid enters into the suction nozzle at Point 1 and flows toward the impeller eye at Point 2. The blades of the impeller trap and accelerate the fluid velocity at Point 3. As the fluid leaves the impeller, its velocity approaches the tip speed of the impeller blades. The volute at Point 4 is shaped like an ever increasing spiral. When the liquid moves at high speed from the close tolerance in the blades to the open spiral volute channel with its ever increasing area, the velocity energy of the liquid is converted into head or pressure energy. With the fluid accumulating its highest pressures at Points 4, the cutwater then directs the fluid to the discharge nozzle at Point 5.

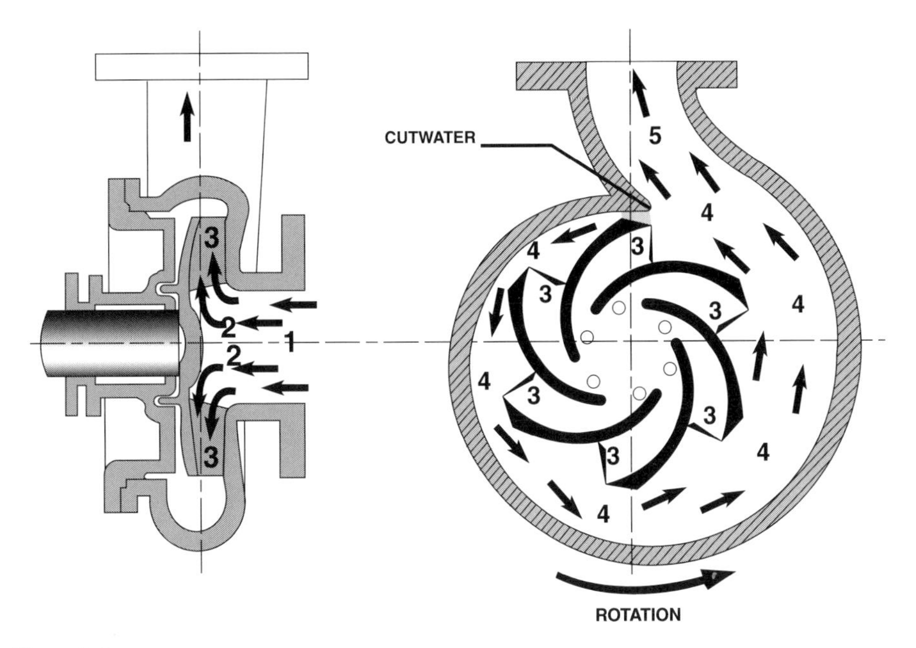

Figure 6–7

Types of centrifugal pumps

Within the large family of centrifugal pumps are smaller groups recognized by the following characteristics:

Overhung impeller

In this group, the impeller or impellers are mounted on the extreme end of the shaft in a cantilevered condition hanging from the support bearings. This group of pumps is also subdivided into a class known as motor-pumps or close-coupled pumps where the impeller is directly mounted onto the motor shaft and supported by the motor's bearings.

Impeller between the bearings

In this group the impeller or impellers are mounted onto the shaft with the bearings on both ends. The impellers are mounted between the bearings. These types of pumps are further divided into single stage (one impeller) or multi-stage pumps (multiple impellers).

Turbine pumps

This group is characterized as having bearings lubricated with the pumped liquid. These pumps are popular in multi-stage construction. The impellers discharge into a vertical support column housing the rotating shaft. These pumps are often installed into deep well water applications. The impellers are commonly mixed flow types, where one stage feeds the next stage through a bell shaped vertical diffuser.

Specific duty pumps

Along with the previously described mechanical configurations, there are some unique types of pumps classified by some special function. Examples are:

- Wastewater pumps have anti-clog impellers to handle large irregular solids.
- Abrasive pumps are made of hardened metal or even rubber-lined to handle abrasive particles in high quantities with minimal erosion.
- Hot water re-circulation pumps are small fractional horsepower models used to heat homes and buildings with circulated hot water through radiators.
- Canned motor pumps are hermetically sealed to prevent emissions, leakage and motor damage. They require no conventional mechanical seal.

The typical ANSI pump

This type of end suction vertically split pump is used extensively in the chemical process industry. It is probably the most popular of all pump designs, see Figure 6–8.

ANSI is an acronym meaning American National Standards Institute. It was previously known as AVS or American Voluntary Standards.

- This pump is most popular in the chemical process industry.
- It can handle abrasive and corrosive liquids.
- It is a one stage, end suction, back-pullout design pump. There is more information on this pump in Chapter 7 about pump curves.
- This pump is available in a wide variety of materials.
- Numerous optional impellers are available.

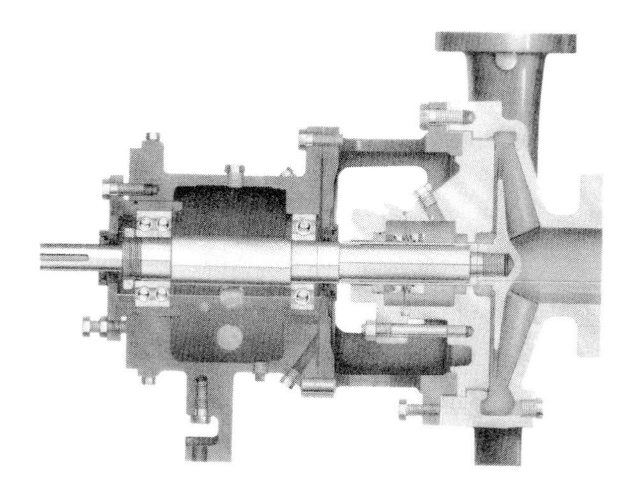

Figure 6–8

API (American Petroleum Institute) pumps

This pump is used extensively in the petroleum industry. This design is similar and yet different from ANSI pumps. It's designed for non-corrosive liquids in applications with high temperature and pressure. It incorporates closed impellers with balance holes (Figure 6–9).

- Complies with API Standard 610.

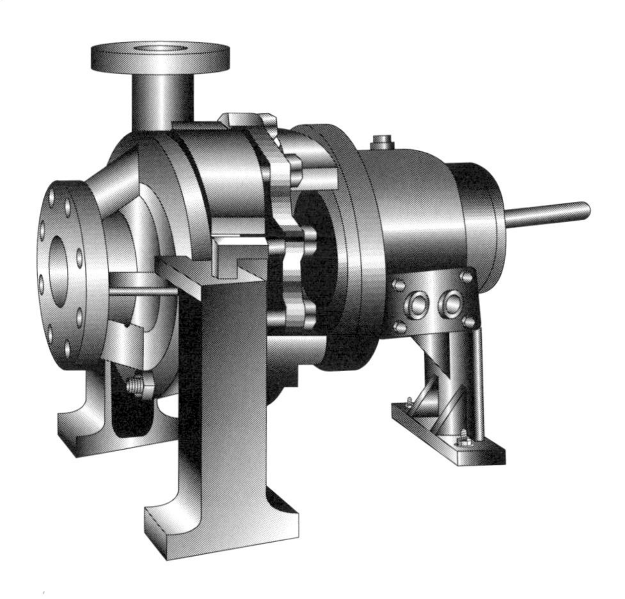

Figure 6–9

- Made for heavy duty, high temperature and pressure applications.
- One stage, end suction, back-pullout construction. See more information on this in Chapter 7.
- The pump's weight and foot supports are mounted on the shaft centerline.
  - Designed for high temperature services above 350° F.
  - This minimizes pipe strain and thermal expansion and distortion.
- Utilizes a closed impeller with balance holes.
- The holes reduce stuffing box pressures and balance axial loading.

Vertical turbine pumps

This type of pump is similar to others except that the impellers discharge into a diffuser bell type housing instead of the volute. The diffuser has multiple veins or ribs that direct the pumped liquid through a column or into the next impeller (Figure 6–10).

- The diffuser equalizes radial loading on the shaft, impeller and journal bearings.

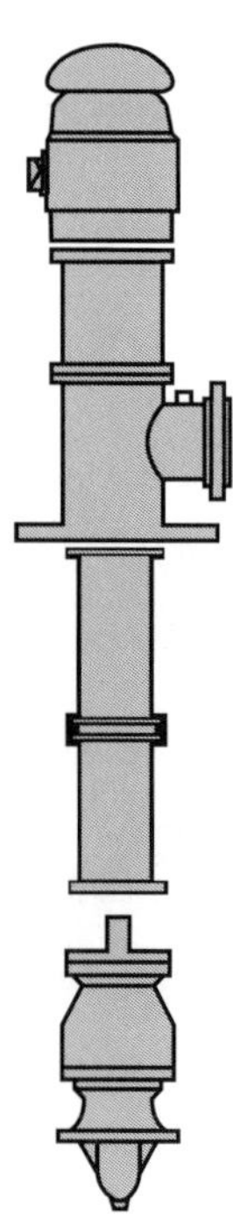

Figure 6–10

- This pump is used where a liquid must be pumped up from subterraneous wells or rivers, or from any open body of fluid (lakes, cooling ponds, tanks and sumps).
- Barrel or canned vertical turbine pumps can be used in-line (Piping, auxiliary booster, and low NPSH applications).
- These pumps don't need priming because the impellers and bell housings are submerged.
- This pump is versatile and adaptable to different applications. It can handle a variety of extension sections depending on the depth of the liquid source.
- The pressure or discharge head is varied by adding and changing stages (impellers).
- These pumps normally use sleeve bearings lubricated with oil, grease, or even the pumped liquid (except for abrasives).
- The motor carries and supports any axial thrust by the pump (hydraulic or mechanical). The motor on these pumps can be fitting with precision rolling element bearings, either angular contact, spherical rollers, or pillow blocks inclined depending on the thrust load and velocity of the shaft.

Non-metallic pumps

This type of pump is used to handle abrasive, chemically corrosive, and oxidizing liquids, where conventional pumps would require exotic alloys. The wet end of these pumps is non-metallic or lined and coated, sealing and isolating any metal component. The power end is normal.

Non-Metallic Construction

- Epoxy Resin
- Phenolic Resin
- Polyester
- Ceramic
- Carbon/Graphite

Lined/Coated Metallic wet parts

- PTFE
- Rubber Lined/Coated
- Glass Lined
- Plastic

- Most of these pump designs are back-pullout construction.
- Some meet complete ANSI specs.

Magnetic drive pumps

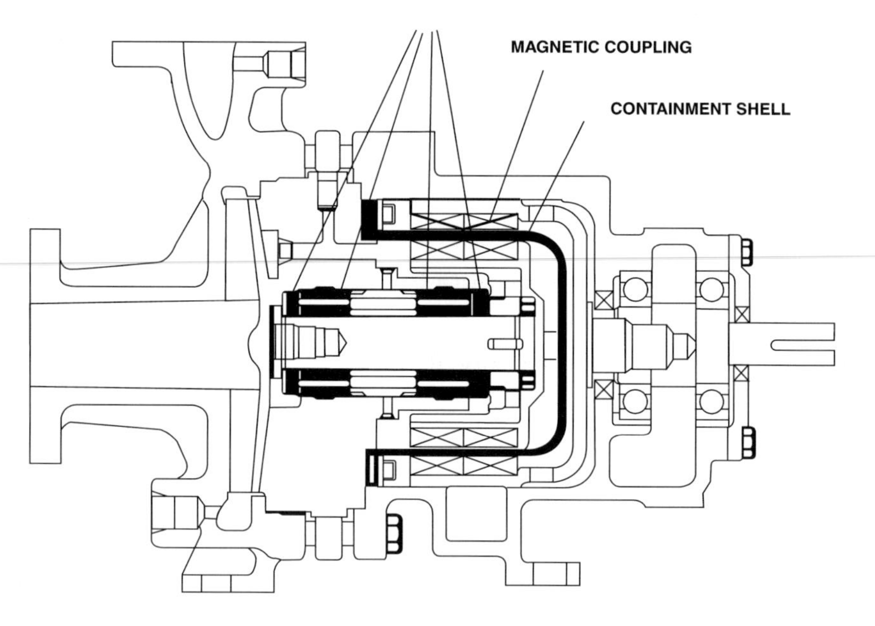

Figure 6-11

This type of pump utilizes a conventional electric motor that drives a set of magnets that drive other magnets fixed to the pump shaft. A non-magnetic housing that isolates the pumped liquid from the environment separates the rotating magnet sets. The impeller, the driven magnet set, shaft and bearing assembly all operate inside the pumped liquid. There are two types of magnet drives:

- Eddy Current electromagnets that can experience some slip inside the pump and may decouple.
- Rare earth permanent magnets with no slip and not subject to decoupling.

The advantages

- No mechanical seal.
- Considered leak proof (although some models use gaskets and o-rings as secondary seals).
- No product loss.
- No exposure (neither liquid nor gaseous) to workers or the environment.

- Considered more reliable than canned motor pumps (the containment shell is larger).

The disadvantages

- Higher initial cost and repair cost.
- Not good at handling abrasives.
- Must be operated very close to the BEP on the curve.
- Cannot run dry.
- Cannot resist extended cavitation.
- The magnets can decouple (requiring stopping and restarting the pump).
- Not as efficient as conventional pumps.
- May require larger motors.
- Tends to heat the pumped liquid.

Canned motor pumps

These pumps incorporate an electric motor whose rotary assembly is hermetically sealed inside a stainless steel or exotic alloy can. The motor, pump shaft and bearings all operate wet inside the pumped liquid as shown in Figure 6–12.

The advantages

- Requires no mechanical seal.
- Only two bearings.
- No product loss.
- Considered leak proof.
- No product exposure to workers or the environment (although some owners manuals offer instructions on what to do 'in case of breach-of-containment'.

The disadvantages

- Higher initial cost and repair cost.
- Fine abrasives will damage the bearings.
- Must be operated very close to the BEP on the curve.
- Cannot run dry or under cavitation.

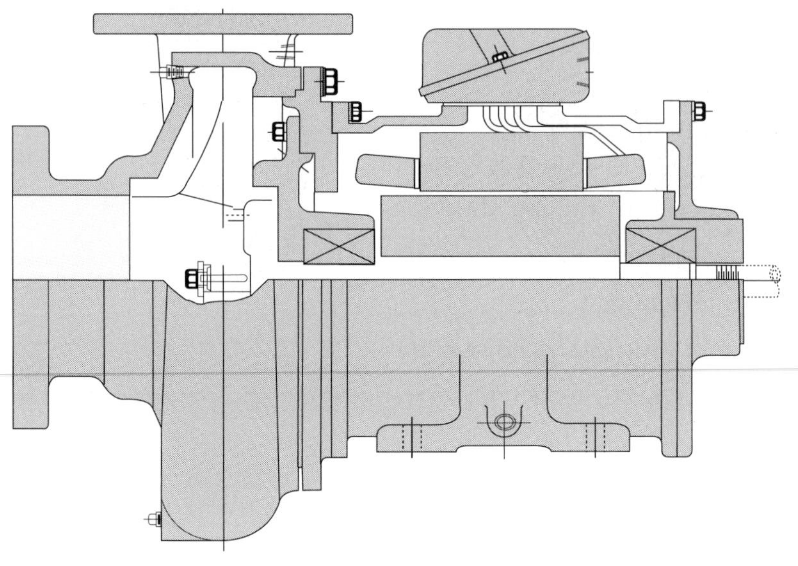

Figure 6–12

- The can may fracture (see advantages).
- Less efficient than conventional pumps.
- May consume more energy (BHP) than conventional pumps.
- Cannot see the direction of rotation.

Pump impellers

The pump impeller receives the pumped liquid and imparts velocity to it with help from the electric motor, or driver. The impeller itself looks like a modified boat or airplane propeller. Actually, boat propellers are axial flow impellers. Airplane propellers are axial flow impellers also, except that they are adapted to handle air.

As a general rule, the velocity (speed) of the impeller and the diameter of the impeller, will determine the head or pressure that the pump can generate. As a general rule, the velocity and the height of the impeller blades, will determine the flow (gpm) that the pump can generate (Figure 6–13).

Remember that pumps don't actually generate flow (no pump in the world can convert three gallons per minute at the suction nozzle into four gallons per minute out of the discharge nozzle), but this is the term used in the industry.

Pump impellers have some different design characteristics. Among

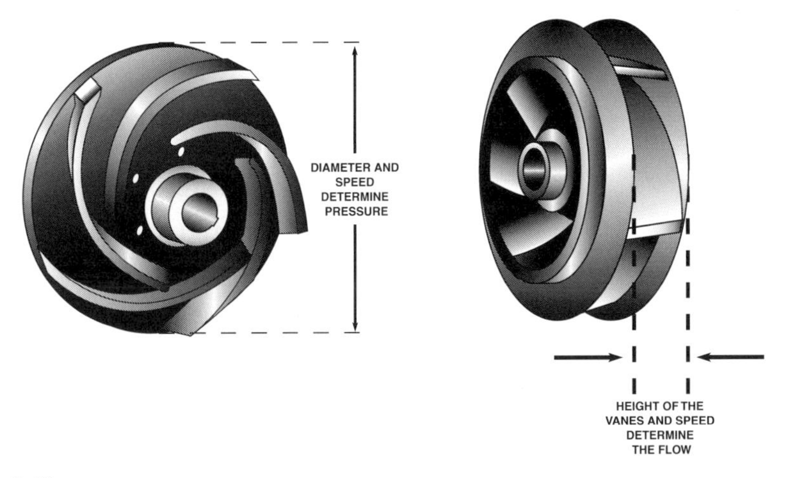

Figure 6–13

them is the way that the impeller receives the liquid from the suction piping. A classic pump impeller receives the liquid at the impeller's ID. By centrifugal force and blade design, the liquid is moved through the blades from the ID to the OD of the impeller where it expels the liquid into the volute channel.

Turbine impellers

On the other hand, turbine impellers receive the liquid at the outside diameter of the impeller, add velocity from the motor, and then expel the liquid, also at the OD to the discharge nozzle. Because these impellers have little available area at the OD, these impellers don't move large quantities of liquid. But, because the liquid's velocity is jerked instantly and violently to a very high speed (remember that a classic centrifugal pump has to **accelerate** the liquid across the blades from the ID to the OD), a lot of energy is added to the fluid and these type pumps are capable of generating a lot of head at a low flow. Additionally, because all the action occurs at the impeller's OD (Remember that there are friction losses and drag as the liquid in a centrifugal pump traverses the impeller blades from ID to OD), there are minimal losses in a turbine pump impeller, which further adds to its high-pressure capacity, see Figure 6–14.

In the case of a **regenerative turbine pump**, any high-energy liquid that doesn't leave the pump through the discharge nozzle is immediately re-circulated back toward the suction where it combines with any new liquid entering into the blades. In this case even more energy is added to already high-energy liquid (thus the name 'regenerative'). This type pump continues to regenerate and compound its pressure or

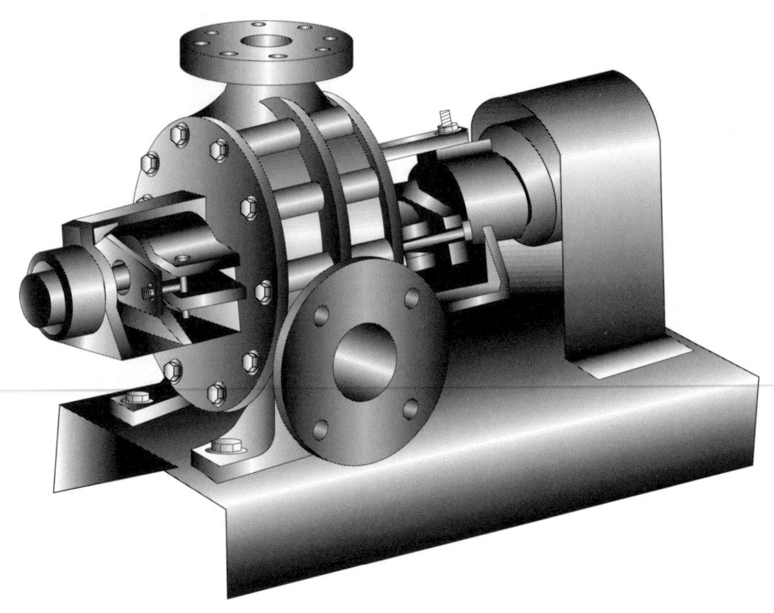

Figure 6–14

discharge head. It makes for a small piece of iron that packs an amazing punch. Regenerative turbine pumps are found on industrial high-pressure washers and enjoy a well-earned reputation as a feed water pump on package boilers.

Conventional impellers

However, most conventional pump impellers receive the fluid into the impeller eye, at the center or inside diameter of the impeller. There are single suction impellers, and dual or double suction impellers with two eyes, one on each side. Dual suction impellers are mostly specified for low NPSH applications because the eye area is doubled (it can receive twice as much fluid at a lower velocity head). Dual suction impellers are mostly found on split case pumps where the shaft passes completely through the impeller. But they can also be found mounted onto the end of the shaft in some special pump designs.

Suction specific speed, Nss

The way that a pump receives the liquid into the impeller determines the available combination of discharge flow and head that the pump can generate. Essentially, it determines the operating window of the pump.

This operating window is quantified or rated by the term 'Suction Specific Speed, Nss'. The Nss is calculated with three parameters, the speed, the flow rate, and the NPSHr. These numbers come from the pump's performance curve, discussed in Chapter 7. The formula is the following:

$$Nss = \frac{N \times \sqrt{Q}}{NPSH^{3/4}}$$

Where: **N** = the speed of the pump/motor in revolutions per minute **Q** = the square root of the flow in gallons per minute at the Best Efficiency Point BEP. For double suction pumps, use ½ BEP Flow. **NPSHr** = the net positive suction head required by the pump at the BEP.

AUTHOR'S NOTE

For the purposes of understanding this concept and formula, there's nothing mathematically significant about the square root of the flow, or the NPSHr to the $^3/_4$ power. These mathematical manipulations simply give us Nss values that are easily understood and recognizable. For example, the health inspector might judge a restaurant's cleanliness on a scale from 1 to 100. We might ask you to rate this book on a scale from 1 to 10. Those are easy numbers to deal with. How would you rate this book on a scale from 2,369 to 26,426,851? This doesn't make sense. Likewise, the mathematical manipulations in the Nss formula serve simply to convert weird values into a scale from 1,000 to 20,000 that cover most impellers and pumps. Values at 1,000 and 20,000 are on the outer fringes. Most pumps register an Nss between 7,000 and 14,000 on a relative scale that is easily understood and comparable to other Nss values of competing pumps, similar pumps, and totally different pumps.

The Nss value is a dimensionless number relating the speed, flow and NPSHr into an operating window that can be expected from a pump. It is an index or goal used by pump design engineers. Consulting engineers use the Nss when comparing similar pumps for correct selection into an application. Once the pump is installed, it becomes a valuable tool for the process engineer, and for the operators interested in keeping the pump running without problems. The Nss is an indication of the pump's ability to operate away from its design point, called the BEP, without damaging the pump.

The Nss value is really simple, although often it is made to appear complicated. The Nss is an equation with a numerator and a denominator. The Nss value is obtained by dividing the numerator by the denominator.

AUTHOR'S NOTE

The operator of a car would know the limits of his automobile. He would or should know if the car is capable of operating safely before launching out on a cross-country trip at highway speeds. He should know how much weight the car can carry safely in the trunk. He should have a general idea if he's getting the expected gasoline mileage from his car. Right? Likewise, the process engineer (and operators) of an industrial pump should know the operating window of the pump. RIGHT?

In the numerator we have the speed and the flow. If we were comparing similar pumps into an application, these multiplied numbers would mostly be a constant. In the denominator we have the NPSHr of the pump (or competing pumps under comparison for an application). As the NPSHr of the pump goes down, the Nss value rises. As the Nss value increases, the operating window of the pump narrows.

Some pump companies will promote and tout their low Nss values. Sometimes a specification engineer will establish a maximum Nss limit for quoted pumps. Let's consider these examples of operating parameters of pumps, and determine the Nss. These values are lifted from the pump performance curves at the BEP.

Parameters	Example 1	Example 2	Example 3
Centrifugal Pump Type/ Liquid	End Suction pump, Single Stage, ANSI Spec/ Cooling Water	End Suction, Single Stage, API # 610 Spec/Kerosene	Dual Suction Impeller, Single Stage, NFPA Code/ Firewater
Pump/Motor Speed	1,750 rpm.	3,500 rpm.	1,780 rpm.
Flow	600 gpm.	1,200 gpm.	4,500 gpm.
NPSHr a BEP.	7 feet	30 feet	20 feet
Nss	$\frac{1750 \times \sqrt{600}}{7^{3/4}} = 9{,}961$	$\frac{3500 \times \sqrt{1200}}{30^{3/4}} = 9{,}458$	$\frac{1780 \times \sqrt{2250}}{20^{3/4}} = 8{,}928$

By using these Nss values, we can interpret the Nss Graph, and get a picture of the operating window of these three pumps. To interpret the graph we start on the left column at the flow in gpm. In Figure 6–15, we draw a line from the flow to the Nss value of the pump, and then reference downward for water, or upward for hydrocarbons.

For the first example, the line terminates at 42%. This means DO NOT

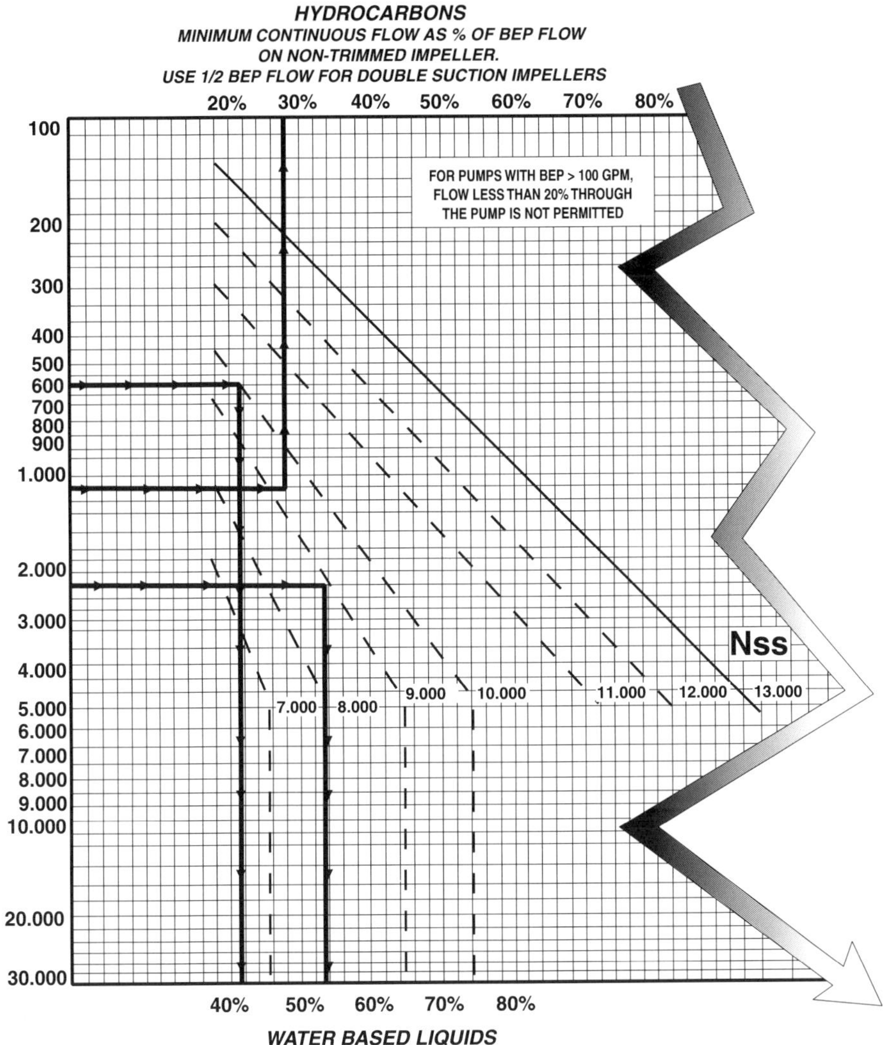

Figure 6–15

operate this pump at less than 42% of the BEP. 42% of 600 gpm is 252 gpm. The operator of this pump should not throttle a control valve and restrict this pump at less than 252 gpm. If the operator throttles this pump to 240 gpm, and goes to lunch, he'll probably have an emergency when he returns from his lunch break. Actually this failure would be an operation-induced failure. If you're mistreating your car, you cannot blame the mechanic.

In the Second example, the line terminates at 29%. This means DO NOT operate this pump at less than 29% of the BEP. 29% of 1200 gpm is 348 gpm. The process engineer should instruct the operators to always maintain the flow above 350 gpm unless he's prepared for pump failure and stalled production.

In the third example, the line terminates at 53%. This means DO NOT run this pump at less than 53% of the BEP. 53% of 4500 gpm is 2385 gpm. Because this is a firewater pump and because firemen need to throttle the nozzles on their fire hoses, then we need to install a pressure relief valve on this system with a discharge bypass line so that the pump dumps the restricted water (less than 2400 gpm) back into the suction tank or lake. If not, this firewater pump is likely to suffer bearing failure during an emergency.

The operating window is the effective zone around the BEP on the pump curve that must be respected by the process engineer and/or the operators of the pump. How far away from the BEP a pump can operate on its performance curve without damage is determined by its impellers suction specific speed.

Open impellers

Impellers are also classified as to whether they are:

1. Totally open,
2. Semi-open (also called Semi-enclosed), and
3. Totally enclosed.

Most totally open impellers are found on axial flow pumps.

This type of impeller would be used in a somewhat conventional appearing pump to perform a chopping, grinding, or macerating action

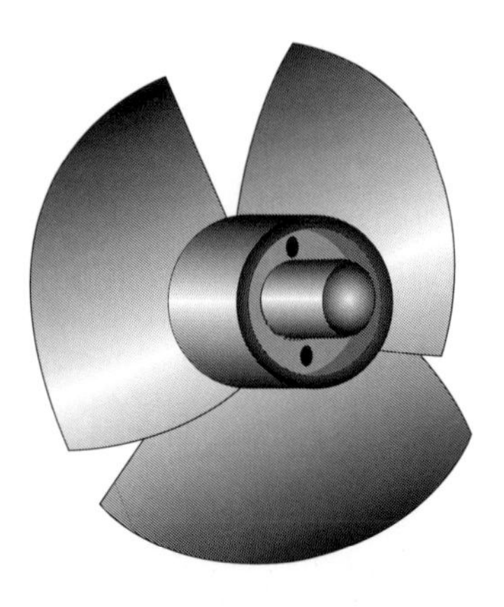

Figure 6-16

on the liquid. The blade in the bottom of the kitchen blender is a macerating axial flow totally open impeller. The totally open axial flow impeller moves a lot of volume flow (gpm), but not a lot of head or pressure. With its open tolerances for moving and grinding solids, they are generally not high efficiency devices.

Semi open impeller

A semi-open impeller has exposed blades, but with a support plate or shroud on one side. Some people prefer the name semi-enclosed. These types of impeller are generally used for liquids with a small percentage of solid particles like sediment from the bottom of a tank or river, or crystals mixed with the liquid (Figure 6–17).

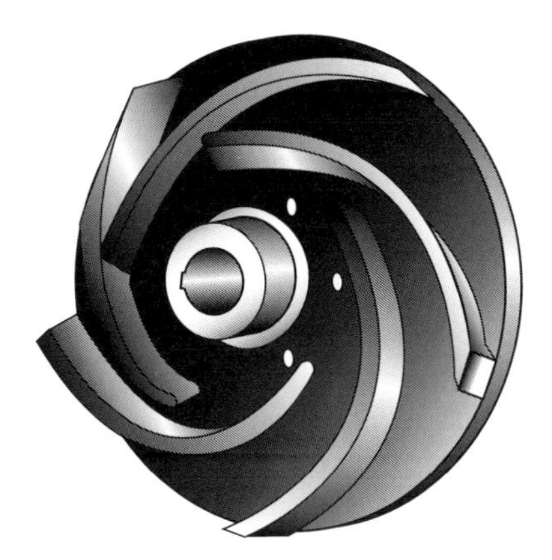

Figure 6–17

The efficiency of these impellers is governed by the limited free space or tolerance between the front leading edge of the blades and the internal pump housing wall. Some pumps have a micrometer gauged jack bolt arrangement on the axial bearing for performing an impeller setting. The impeller setting corrects for erosion wear and thermal expansion in this tight tolerance, returning the pump to its original efficiency.

Totally enclosed impeller

Totally enclosed impellers are designed with the blades between two support shrouds or plates. These impellers are for totally clean liquids because tolerances are tight at the eye and the housing, and there is no room for suspended solids, crystals or sediment, see Figure 6–18.

Figure 6–18

Solid contamination will destroy the tolerance between the OD of the eye and the bore of the pump housing.

This specific tolerance governs the efficiency of the pump.

The tolerance between the OD of the impeller eye and the internal bore of the pump housing is set at the factory based on the temperature of the application and thermal growth of the pump metallurgy. This tolerance tends to open with time for a number of reasons. Among them: erosion due to the passage of fluid, the lubricating nature of the liquid, suspended solids and sediment will accelerate the wear, cavitation damage, play in the bearings, bent shafts and unbalanced rotary assemblies, and any hydraulic side loading on the shaft and impeller assembly.

Wear bands

Some pump companies will design replaceable wear bands for the OD of the impeller eye and the bore of the pump housing. It's said that the pump loses 1.5% to 2% efficiency points for every one thousandths wear in a wear band beyond the factory setting. Therefore, by changing wear bands, the pump is returned to its original efficiency. Because of this, the term wear band is a misnomer. A better term would be 'efficiency band' (Figure 6–19).

The replaceable wear bands can also be made in a machine shop in a pump maintenance function. It is important that the new wear band material is made of a non-galling, and non-sparking material softer than the pump housing metallurgy. Plastic, composite, fiberglass and carbon graphite wear band are perfectly good. Be sure the material is compatible with the pump's metallurgy and the pumped liquid. It's not

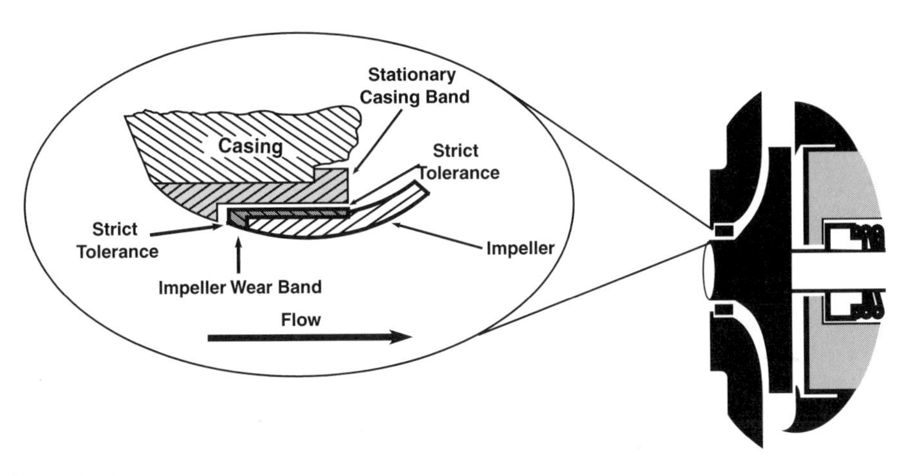

Figure 6-19

necessary that they be made of metal. Remember that their function is not to wear, but to control the tolerance and efficiency of the pump.

Specific speed, Ns

Another distinction in impellers is the way the liquid traverses and leaves the impeller blades. This is called the Specific Speed, Ns. It is another index used by pump designers to describe the geometry of the impeller and to classify impellers according to their design type and application. By definition, the Specific Speed, Ns is the revolutions per minute (rpm) at which a geometrically similar impeller would run if it were of such a size as to discharge one gallon per minute at one foot of head.

The equation for determining the Ns is similar to equation for the Nss, except that it substitutes the NPSHr in the denominator with the pump's discharge head:

$$Ns = \frac{N \times \sqrt{Q}}{H^{3/4}}$$

Where: N = the speed of the pump/motor in revolutions per minute
Q = the square root of the flow in gallons per minute at the Best Efficiency Point BEP. For double suction impellers, use ½ BEP flow.
H = the discharge head of the pump at the BEP.

The Specific Speed is a dimensionless number using the formula above. Pump design engineers consider the Ns a valuable tool in the development of impellers. It is also a key index in determining if the pump

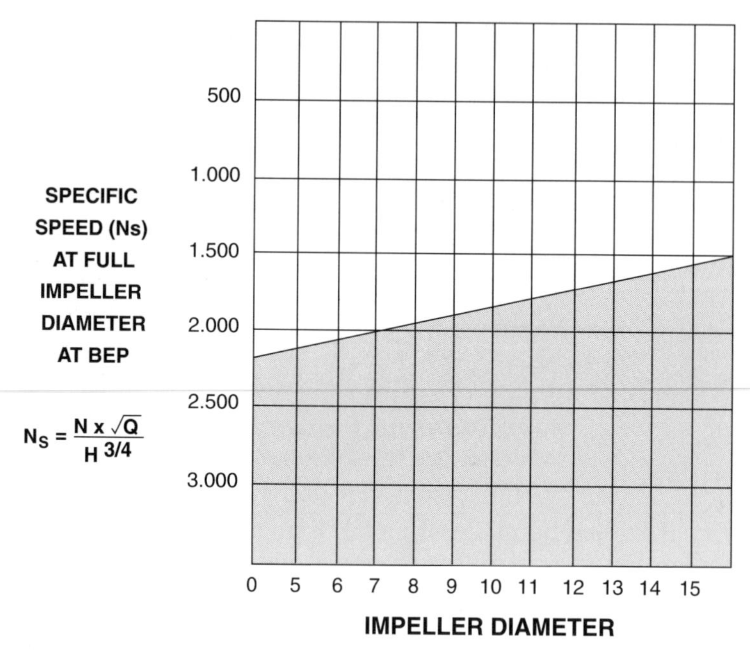

Figure 6–20

should be specified with the single volute designed casing, or the double volute designed casing (Figure 6–20).

Some pumps are operated at or close to their best efficiency points. Other pumps must run far to the left or right of their best efficiency

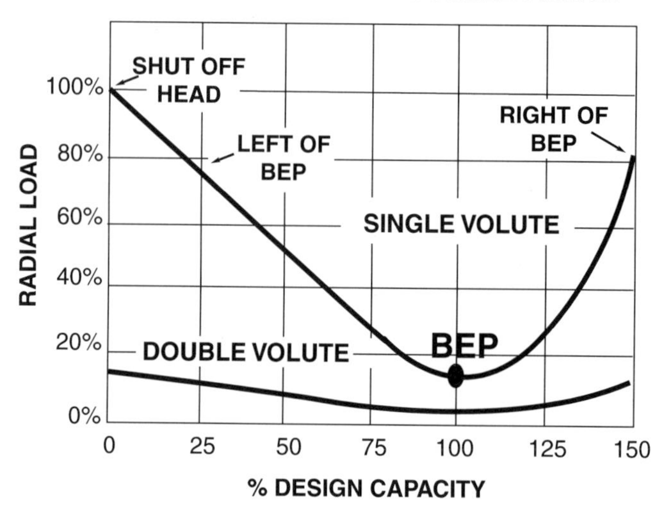

Figure 6–21

points. Pumps operating away from their best efficiency points tend to develop hydraulic side loads that can stress the shaft, damaging the bearings, wear bands, and mechanical seal (Figure 6–21).

There is more information on this in Chapter 9. Dual volute casings tend to equalize the radial hydraulic forces around the pump impeller, thus expanding the operating window of the pump. The Ns is a guide in selecting the adequate volute design.

The Ns is useful in analyzing a problematic pump and in purchasing a new pump. When the parameters of a new pump are determined, the speed, flow, and head can be worked through the Ns formula to give a value indicating a certain type impeller design. See Figure 6–22.

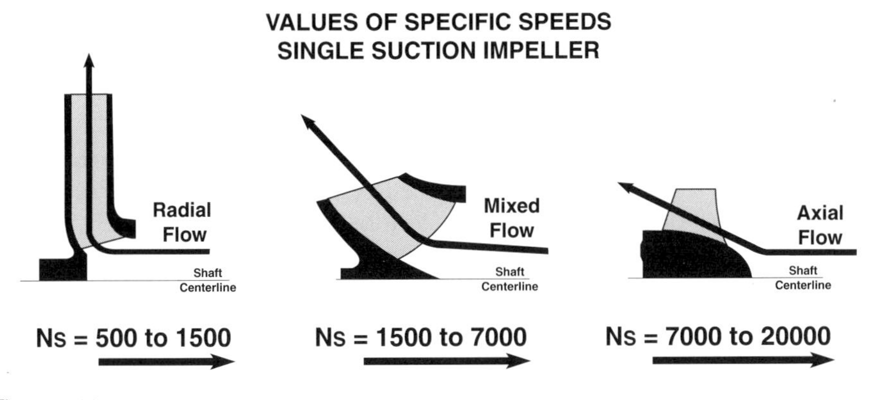

Figure 6–22

Pumps should be considered when their impeller profile corresponds to the calculated Ns value.

Radial vane impellers (Ns values between 500 and 1,500) generate head with pure centrifugal action. In Francis and Mixed vane impellers (Ns values between 1,500 and 8,000), some head is developed by centrifugal action and other head is developed by the impeller's design. These impellers are popular in multi-stage vertical turbine pumps. Also with these designs, the wider impellers vanes indicate that these pumps are better with developing flow and not so much head. Axial flow impellers (Ns above 8,000) are almost exclusively specified in high flow applications with little head.

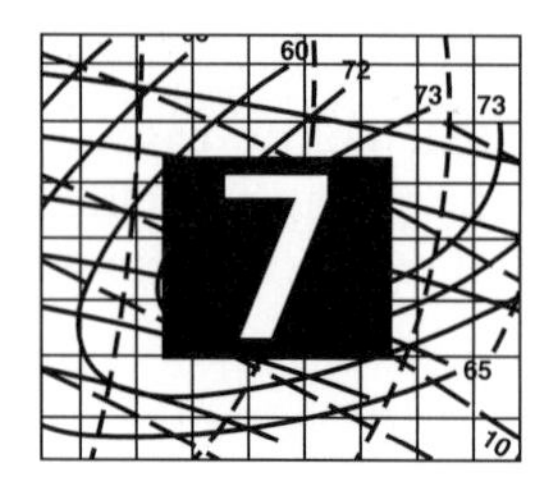

Understanding Pump Curves

Pump performance curves

Pump performance curves are the least used, least consulted, least appreciated, and least understood aspect of the world of industrial pumps. The plant personnel who most need their pump curves, mechanics and operators, generally don't have the curves and accompanying information at their disposal. The people who control the performance curves store them in a file, in a drawer, in a cabinet that's almost never opened. They don't share the information contained in the curves with the people who need it. Maybe it's because they themselves don't understand the information to share it. In the next few paragraphs and pages, we're going to explain the pump performance curves. This might be the most important chapter of the book.

In reality, the performance curve is easy to understand. It isn't rocket science. The performance curve indicates that the pump will discharge a certain volume or flow (gpm) of a liquid, at a certain pressure or head (H), at an indicated velocity or speed, while consuming a specific quantity of horsepower (BHP). The performance curve is actually four curves relating with each other on a common graph. These four curves are:

1. The Head-Flow Curve. It is called the H-Q Curve.
2. The Efficiency Curve.
3. The Energy Curve. It records Brake Horsepower, BHP.
4. The Pump's Minimum Requirement Curve. Its called Net Positive Suction Head required, NPSHr.

Think of the pump curve like the dashboard or control panel of a car. No one would operate a car without the dash instrumentation panel.

The information on the dash panel is located right in front of the eyes of the operator of the car. It's a shame that most pump operators don't have their control panel (the curve) before their eyes, or even within reach, as they operate the pumps. This is the source of many problems with pumps.

History

Some three thousand years ago, the ancient Romans and Greeks understood the hydraulic laws that govern today's modern pumps. They had already calculated the physics and math required to bring water from the mountain streams, down through giant aqueducts and underground clay pipes, and spray a stream of water 12 ft up into the air in the fountain at the public square. They understood the laws of gravity and the concept of atmospheric pressure. They knew at what volume, and at what speed, the water had to fall through the troughs in the aqueducts, to arrive into the heart of the cities and supply the needs of the growing population.

About 2,200 years ago, a Grecian named Archimedes, developed the first practical pump. He took a hollow tree trunk, and carved an internal spiral corkscrew type groove from one end of the trunk to the other. By lowering one end of the tree trunk into a mountain lake and rotating the trunk (on its axial centerline), the water flowed upward through the spiral groove and dropped out of the upper end of the tree trunk. By positioning the upper end of the tree trunk over a trough of an aqueduct, the water began flowing down the aqueduct to irrigate crops, or to supply the city below with fresh water.

In those days, there were no oil refineries, nor bottlers of carbonated soda, nor sulfuric acid plants. There was only one liquid to consider, and move in large quantities ... fresh water from the mountains. With only one liquid under consideration, fresh water, and no sophisticated instrumentation, they measured the water's force, or pressure, in terms of elevation. It is for this reason that today all over the world, pump manufacturers use the term 'Head' measured in meters or feet of elevation to express pressure or force. The term 'flow' expresses volume over time, such as gallons per minute, or cubic meters per second.

Head versus pressure

There's a language barrier between the pump manufacturers and the pump users. They use different terminology. Pump users, the operators and mechanics, use pressure gauges that read in psi, pounds per square inch (or kilograms per square centimeter, in the metric system). The pump manufacturer denotes pressure in feet of head (or meters of head). The pump operator needs a pump that generates 20 psi. The manufacturer offers a model that generates 46 ft of head.

To understand pumps and analyze their problems, its necessary to dominate the formula that changes feet of head (H) into psi. This is explained in Chapter 2, but here is a brief review:

The formula is:

$$\textbf{Pressure in psi} = \frac{\textbf{H (Head in feet)} \times \textbf{sp. gr.}}{\textbf{2.31}}$$

And in the other direction:

$$\textbf{Head in Feet} = \frac{\textbf{psi} \times \textbf{2.31}}{\textbf{sp. gr.}}$$

If the liquid is water, the specific gravity is 1.00. We see that two factors separate 'psi' from 'head in feet'. First is the 2.31 conversion factor, and second, the specific gravity.

The pump companies develop their curves using head in feet (H), because when they make a new pump, they don't know the ultimate service of the pump (they don't know the liquid that the pump will be pumping), but they do know how many feet of elevation the pump can raise that liquid. This is why it's necessary to specify pumps in feet of head and not in psi. Let's begin by exploring the H-Q curve of the pump, using feet of head.

H-Q

The matrix of the pump curve graph is the same as the mathematical 'x-y' graph. On the horizontal line, the flow is shown normally in gallons per minute or cubic meters per second. The vertical line shows the head in feet or meters. See Figure 7–1.

By definition, the pump is a machine designed to add energy to a liquid with the purpose of elevating it or moving it through a pipe. The pump can elevate a liquid in a vertical tube up to a point where the weight of the liquid and gravity will permit no more elevation. The energy contained in the liquid's weight is the same as the energy produced by the pump. This point on the pump curve would be the 'shut-off head'. Shut-off head is the point of maximum elevation at zero flow. It's seen in Figure 7–2.

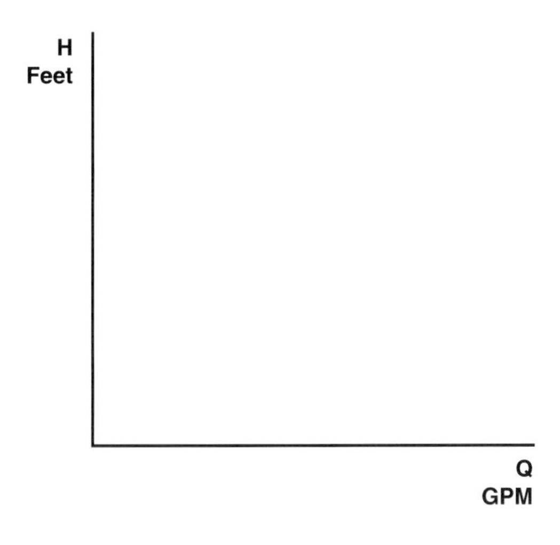

Figure 7–1

Once again, imagine starting a pump and raising the fluid in a vertical tube to the point of maximum elevation. On the curve this would be maximum head at zero flow. Now, rotate the running pump on its centerline 90°, until the vertical tube is now in a horizontal position. The very action of rotating the running pump on its centerline would trace the pump's curve. Any elevation in feet would coincide with a flow in gallons per minute. Consider the graph show in Figure 7–3.

On the graph, if point 'A' represents 10 ft of head at 0-gpm, and if point 'F' represents 10 gpm at 0 ft of head, then point 'C' on the curve represents 8 ft of head at 6-gpm. Here we see that the pump is always on its curve. The pump can operate at any point on this curve from point 'A' to point 'F'. At any specific head, this pump will pump a specific flow, or gpm corresponding to the head.

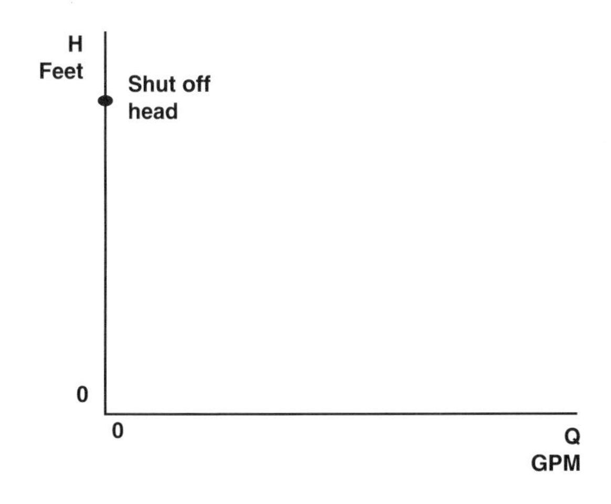

Figure 7–2

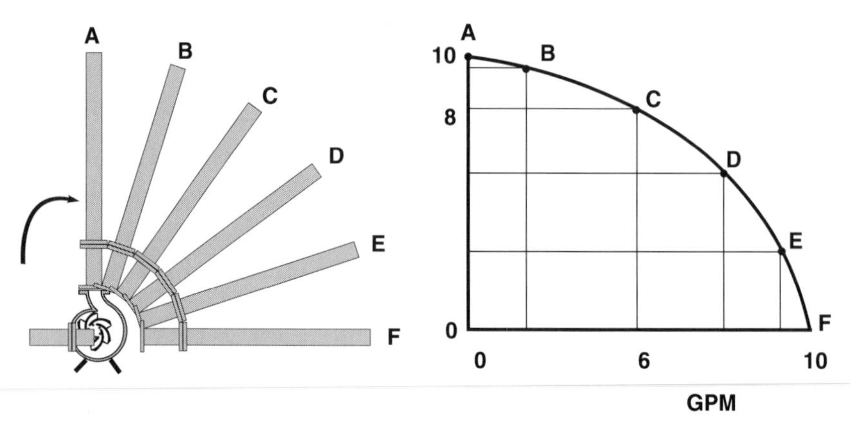

Figure 7–3

AUTHOR'S NOTE

Sometimes you hear people say that the pump is operating off its curve. If the velocity, the impeller diameter and design are correct, if the pump has all its parts installed and functioning correctly, including the mechanical seal and coupling, it is impossible to operate off the curve. The pump will be somewhere on its curve between shut-off head and maximum flow a zero elevation.

The pump can be too far to the right, or too far to the left of its best efficiency point (BEP) but it cannot be off the curve. Conceivably, the pump can be operating off the graph, and even off the page, but it cannot be off the curve. If the pump is off the curve, something else is out of control, like the velocity, or impeller diameter, assembled parts and tolerances. Now, the 'lack of control' is the real problem, and not the pump.

Pump efficiency

Let's talk about the pump efficiency. Imagine a small pump connected to a garden hose squirting a stream of water across the lawn. You could direct the flow from the hose up into the air at about a 45-degree angle, and the stream would arc upward and attain its best distance of reach from the nozzle or launch point. The stream of water would attain a specific height into the air and a specific distance. The efficiency curve of a pump is seen as the trajectory or arc of a stream of water. When squirted from a hose, the elevation that attains the best distance, when plotted onto the pump curve, is called the best efficiency point (BEP). On the pump curve, it is seen as in Figure 7–4.

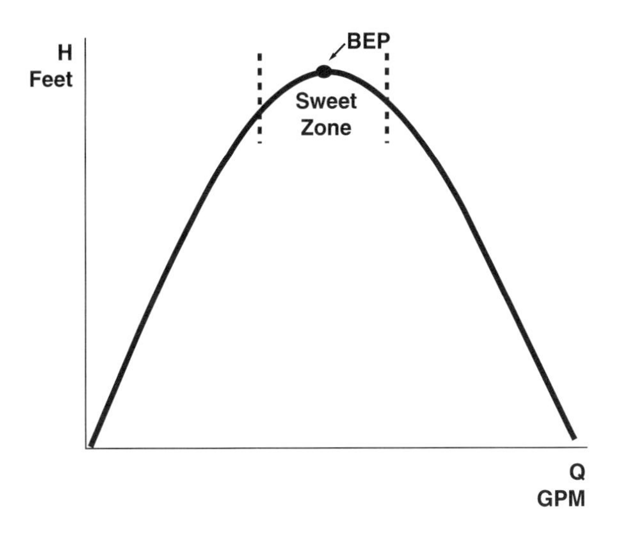

Figure 7–4

The energy (BHp) curve

Next, let's consider the energy curve, the brake horsepower (BHp), required by the pump. This curve is probably the easiest to interpret because it is practically a straight line. Consider the following: the pump consumes a certain quantity of energy just to maintain shut-off head. Then, as flow begins and increases, the horsepower consumption normally increases. (On certain specific duty pumps, the BHp may remain mostly flat or even fall with an increase in flow.) The BHp curve is normally seen this way (Figure 7–5).

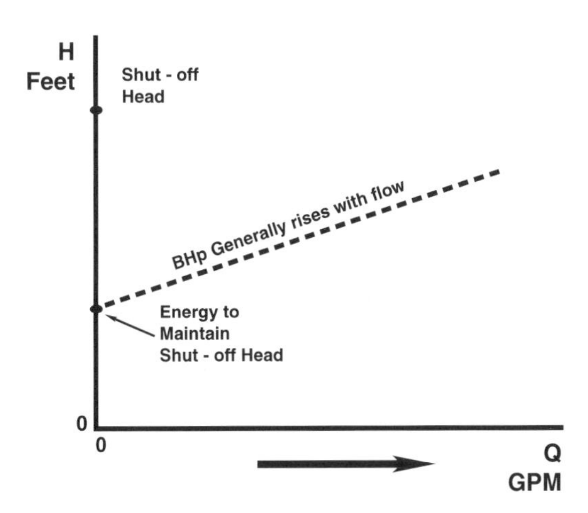

Figure 7–5

The pump's minimum requirements (NPSH)

The last component of the pump performance curve is the curve of the minimum requirements, or NPSH. Actually, the reading on the pump curve is the NPSHr, the Net Positive Suction Head required by the pump. There is a complete discussion on NPSHr and NPSHa, and the result of not respecting or understanding them in Chapters 3 and 4. Basically, the NPSHr curve, beginning at 0 flow, is mostly flat or modestly rising until it crosses through the BEP zone. As the NPSHr curve crosses through the BEP of the pump, the curve and values begin rising exponentially. Normally it is seen this way (Figure 7–6).

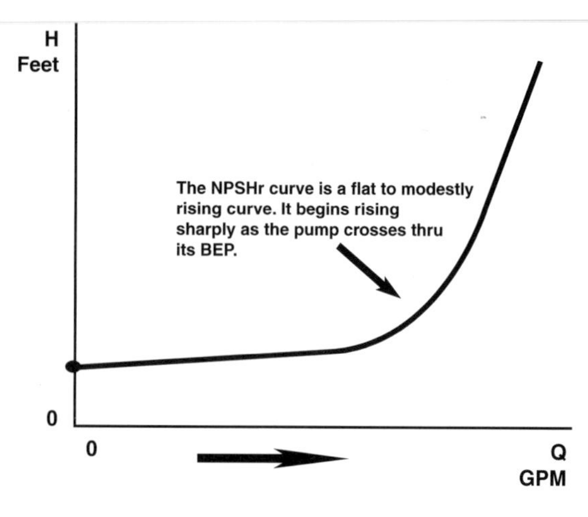

Figure 7–6

Review

See Figure 7–7 for the pump performance curve components.

As you can see in the four components of the pump curve:

- At point 'A' on the H-Q curve, the pump is pumping Q gpm (gallons per minute), at H feet of head. This point on the curve corresponds to the best efficiency, and it is also seen at approximately the middle of the energy curve, and also on the NPSHr curve where it begins its sharp rise.
- At point 'B', the flow is reduced and the head is elevated on the H-Q curve. The pump is being operated to the left of its best efficiency zone. Note that the pump has lost efficiency at this point. The minimum requirements of the pump, the NPSHr, and the horsepower consumption, BHP, have also been reduced, but with the efficiency drop and reduced flow, the pump is vibrating and heating the pumped liquid. The shaft is under deflection, causing stress to the bearings and mechanical seal (or shaft packing rings).

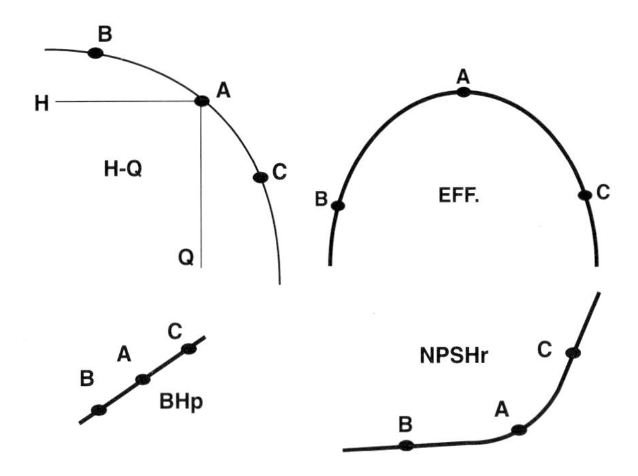

Figure 7–7

- At point 'C', the flow is high and the head (pressure) is low on the H-Q curve. This pump is operating with reduced efficiency, this time to the right of the optimum efficiency point. The BHP is rising and may overload the installed motor. The NPSHr has risen to the point that the pump is being strangled; the liquid is leaving the pump faster than it can come into the pump. The pumped liquid is prone to vaporize or boil. This is the zone where classic vaporization cavitation occurs. And, the shaft is under a deflection load, stressing the seal and bearings.

Let's see these four elements, as they appear on the same graph (Figure 7–8).

You can see in Figure 7–9 that the pump should run at or near zone 'A', its best efficiency point, the BEP. This is the preferred sweet or happy zone. The pump should be specified and operated in this zone.

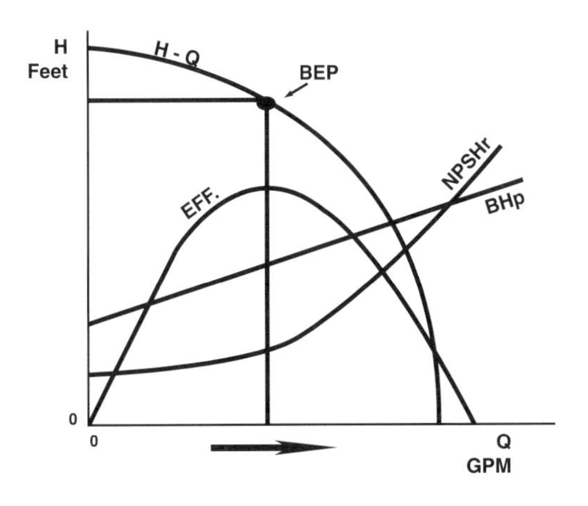

Figure 7–8

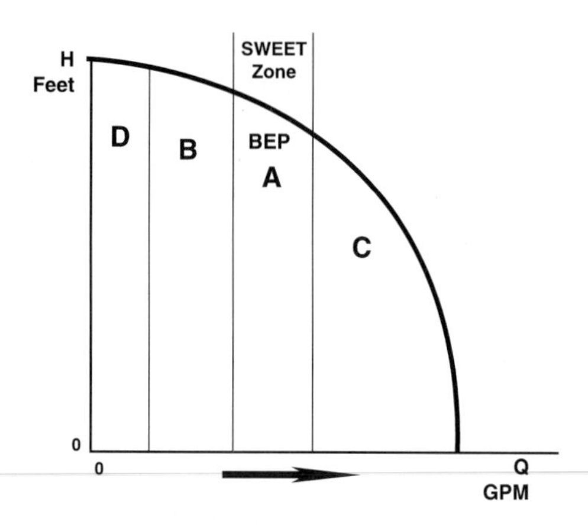

Figure 7–9

Avoid zones 'C' and 'D' at all times. The pump can be operated in zone 'B' only if it is necessary.

Zone 'B' is slightly to the left of the BEP. At this point the pump and impeller is slightly over-designed for the system. The pump will suffer a loss of efficiency. Radial loading is generated on the shaft that can stress the bearings and seal and may even break the shaft. If it is necessary to operate the pump in this part of its curve (to the left of the BEP), for more than a few hours, you should install an impeller with a reduced diameter. Remember that the back pullout pump exists for rapid and frequent impeller changes (see Chapter 6). By reducing the impeller diameter, you can maintain the head and pressure, but at a reduced flow. Figure 7–10 illustrates this point.

In zone 'C' the pump is operating to the right of the BEP and it is inadequately designed for the system in which it is running. To a point

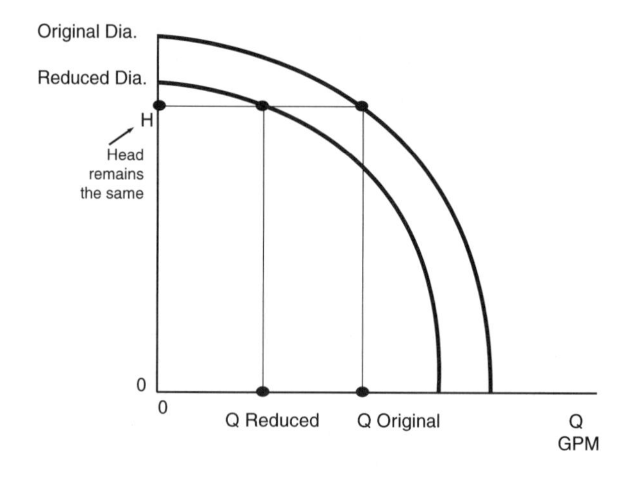

Figure 7–10

you could meet the requirements of flow, but not the requirements of head or pressure. The pump is prone to suffering cavitation, high flow, high BHp consumption, high vibrations, and radial loading (about 240° from the cutwater), resulting in shaft deflection. To counteract these results, the operator should restrict the control valve on the pump's discharge to reduce the flow.

Operating the pump in zone 'D' is very damaging to the pump. Now the pump is severely over-designed for the system, too far to the left of the BEP. The pump is very inefficient with excessive re-circulation of the fluid inside the pump. This low flow condition causes the fluid to overheat. The pump is suffering high head and pressure, and radial loading (about 60° from the cutwater), shaft deflection and high vibrations. To deal with or alleviate these results, you need to modify or change the system on the pump's discharge (ex. reduce friction and resistance losses on the discharge piping), or change the pump (look for a pump whose BEP coincides with the head and flow requirements of the system).

In the final analysis, pumps should be operated at or near their BEP. These pumps will run for years without giving problems. The pump curve is the pump's control panel, and it should be in the hands of the personnel who operate the pumps and understood by them.

Special design pumps

The majority of centrifugal pumps have performance curves with the aforementioned profiles. Of course, special design pumps have curves with variations. For example, positive displacement pumps, multi-stage pumps, regenerative turbine type pumps, and pumps with a high specific speed (Ns) fall outside the norm. But you'll find that the standard pump curve profiles are applicable to about 95% of all pumps in the majority of industrial plants. The important thing is to become familiar with pump curves and know how to interpret the information.

Family curves

At times you'll find that the information is the same, but the presentation of the curves is different. Almost all pump companies publish what are called the 'family of curves'. The pump family curves are probably the most useful for the maintenance engineer and mechanic, the design engineer and purchasing agent. The family curves present the entire performance picture of a pump.

- The family curve shows the range of different impeller diameters that can run inside the pump volute. They're normally presented as various parallel H-Q curves corresponding to smaller diameter impellers.
- Another difference in the family curves is the presentation of the energy requirements with the different impellers. Sometimes the BHp curves appear to be descending with an increase in flow instead of ascending. Sometimes, instead of showing the horsepower consumed, what we see is the standard rating on the motor to be used with this pump. For example, instead of showing 17 horsepower of energy consumed, the family curve may show a 20-horsepower motor, which is the motor you must buy with this pump. No one makes a standard 17 horsepower motor.
- By showing numerous impellers, motors and efficiencies for one pump, the family curve has a lot of information crushed onto one graph. So to simplify the curve, the efficiencies are sometimes shown as concentric circles or ellipses. The concentric ellipses demonstrate the primary, secondary and tertiary efficiency zones. They are most useful for comparing the pump curve with the system curve. (The system curve is presented in Chapter 8.)
- Normally the NPSHr curve doesn't change when shown on the family curve. This is because the NPSHr is based on the impeller eye, which is constant within a particular design, and doesn't normally change with the impeller's outside diameters. In all cases the impeller eye diameter must mate with the suction throat diameter of the pump, in order to receive the energy in the fluid as it comes into the pump through the suction piping.

Figure 7–11 is an example of a family curve for an industrial chemical process pump.

Next, let's consider the family curve for a small drum draining or sump pump. Note that this pump is not very efficient due to its special design. The purpose of this pump is to quickly empty a barrel or drum to the bottom through its bung hole on the top. A typical service would be to mix additives or add treatment chemicals to a tank or cooling tower. This pump can empty a 55 gallon barrel in less than a minute while elevating the liquid to a height of some 25 ft. Observe that the NPSHr doesn't appear on this curve. This is because the NPSHr is incorporated into the design of this specific duty pump. (Remember that it can reach into a drum through the top and drain it down to the bottom.) This is also the reason for the reduced efficiency. Also, notice that the BHp requirements are based on a specific gravity of 1.0 (water). When the liquid is not water, the BHp is adjusted by its

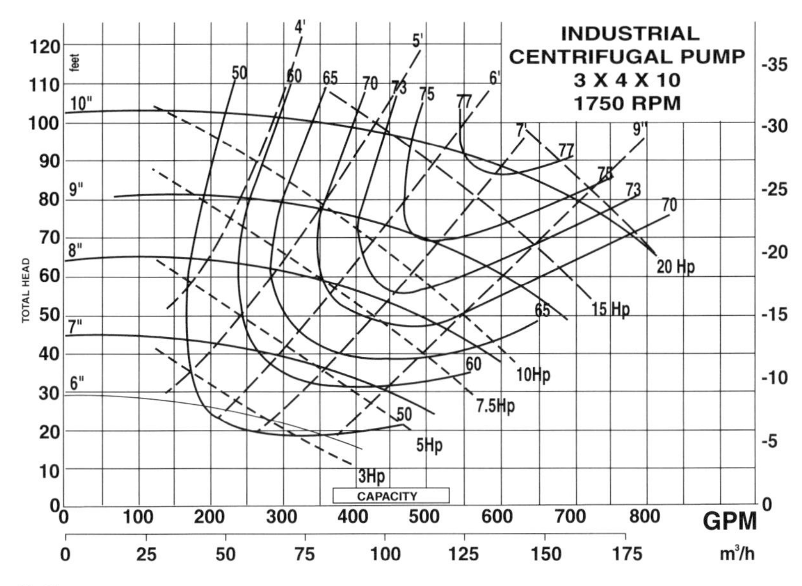

Figure 7–11

specific gravity. Observe that the profiles of this curve are similar to other centrifugal pumps.

See the following curve (Figure 7–12).

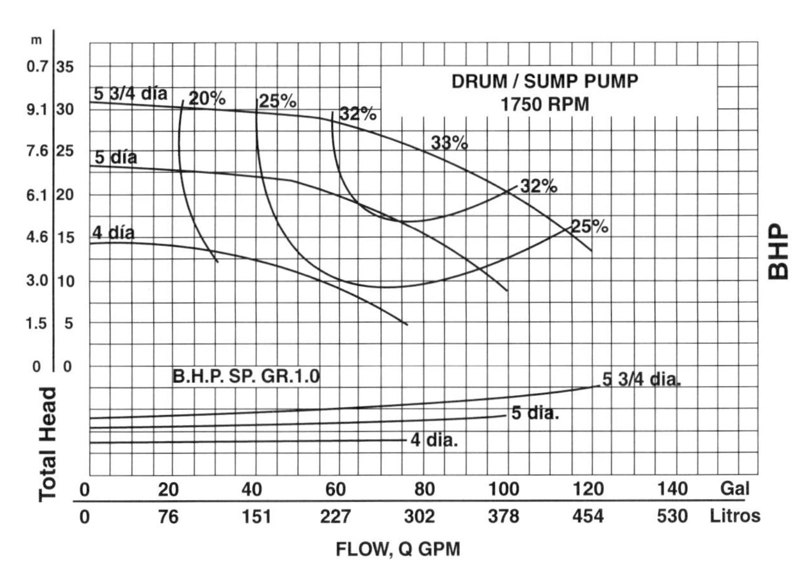

Figure 7–12

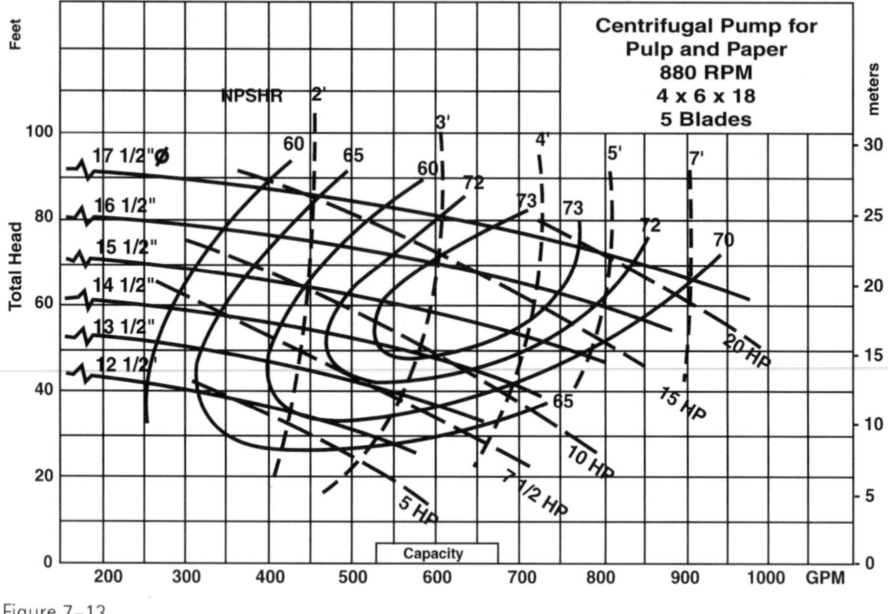

Figure 7–13

Next, consider this family curve for a centrifugal pump used in the pulp and papermaking industry (Figure 7–13).

The next graph is a typical family curve for a firewater pump (Figure 7–14):

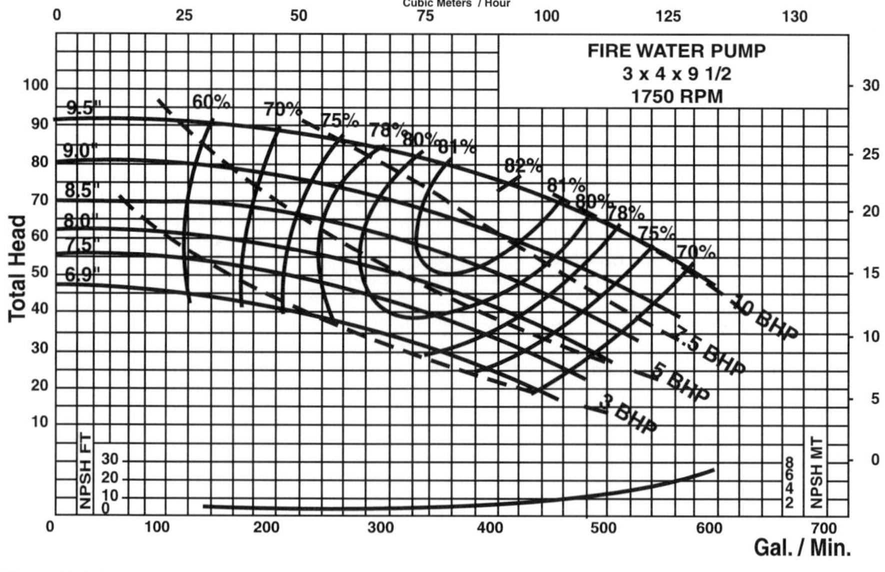

Figure 7–14

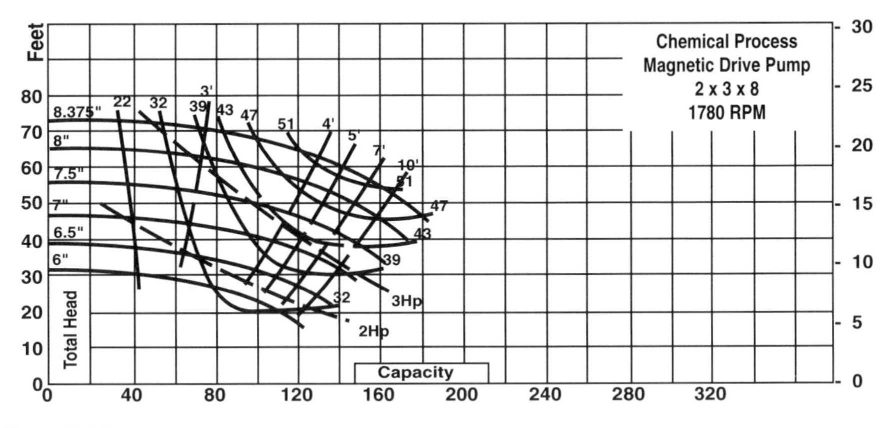

Figure 7–15

Observe this presentation of a family curve for a mag-drive pump used in the chemical process (Figure 7–15).

The graph below is a family curve for a petroleum-refining pump meeting API standards (Figure 7–16).

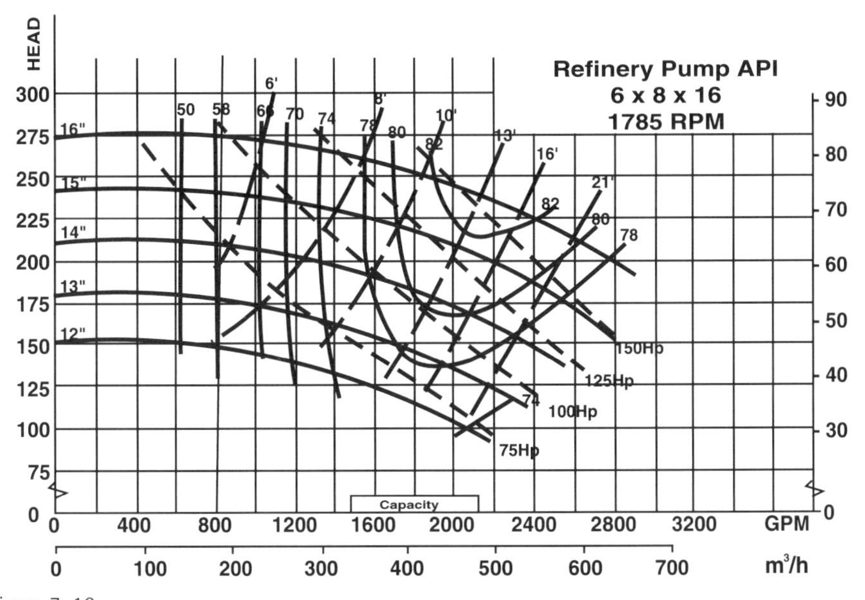

Figure 7–16

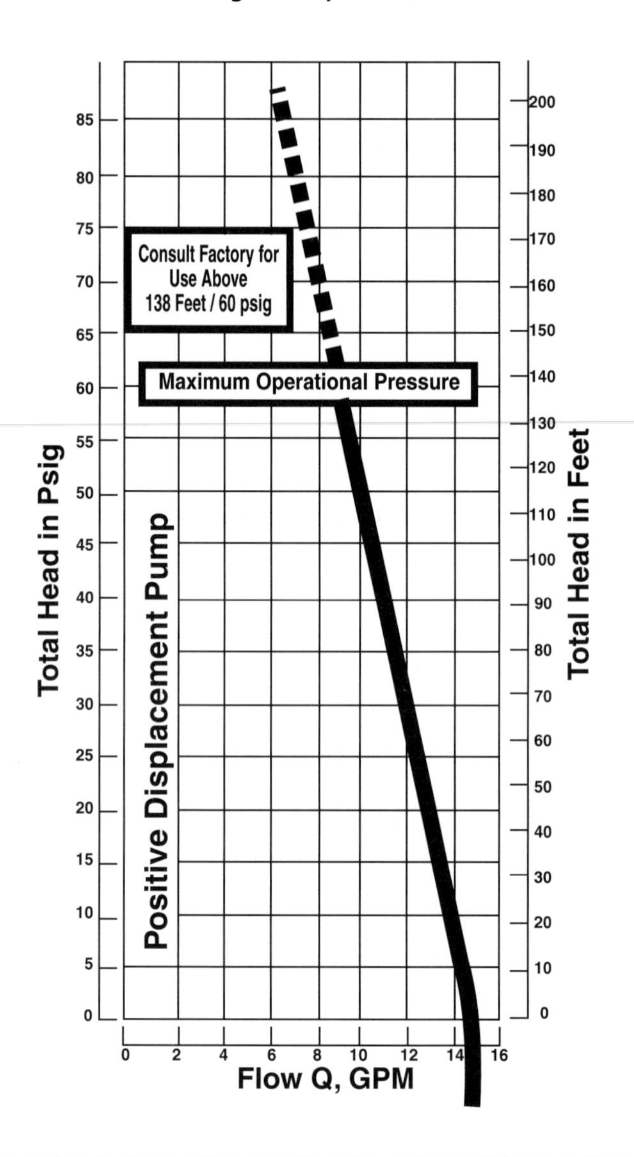

Figure 7–17

Although the pump is not part of this discussion, we present a curve of a positive displacement (PD) pump (Figure 7–17).

On seeing and examining these different pump curves, notice that all curves contrast head and flow. And, in every case the head is decreasing as the flow increases.

Except for the curve of the PD pump, the other pump curves show various diameter impellers that can be used within the pump volute. And, on all these family curves, the efficiencies are seen as concentric ellipses. There is very little variation in the presentation of the BHp and

NPSHr. Notice that the small drum pump doesn't show the NPSHr. This is because this pump, by design, can drain a barrel or sump down to the bottom without causing problems.

To end this discussion, the curve is the control panel of the pump. All operators, mechanics, engineers and anyone involved with the pump should understand the curve and it's elements, and how they relate. With the curve, we can take the differential pressure gauge readings on the pump and understand them. We can use the differential gauge readings to determine if the pump is operating at, or away (to the left or right) from its best efficiency zone and determine if the pump is functioning adequately. We can even visualize the maintenance required for the pump based on its curve location, and visualize the corrective procedures to resolve the maintenance.

Up to this point, you probably didn't understand the crucial importance of the pump curve. With the information provided in this chapter, and this book, we suggest that you immediately locate and begin using your pump curves with suction and discharge gauges on your pumps.

Get the model and serial number from your pumps, and communicate with the factory, or your local pump distributor. They can provide you with an original family curve, and the original specs, design and components from when you bought the pump. A copy of the original family curve is probably in the file pertaining to the purchase of the pump. Go to the purchasing agent's file cabinet.

Nowadays, some pump companies publish their family curves on the Internet. You can request a copy with an e-mail, phone call, fax, or letter. The curves and gauges are the difference between life and death of your pumps. The pump family curve goes hand in hand with the system curve, which we'll cover in the next chapter.

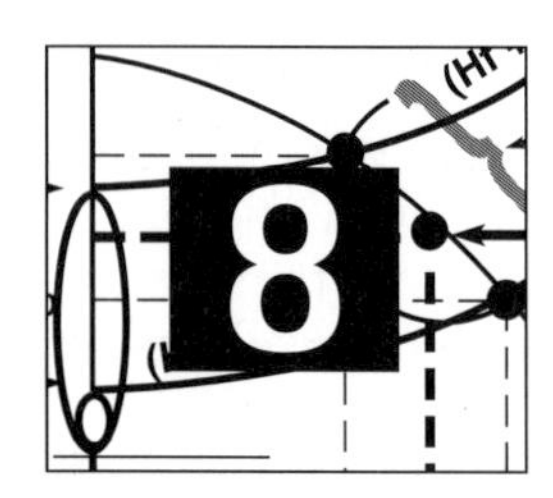

The System Curve

The system controls the pump

All pumps must be designed to comply with or meet the needs of the system. The needs of the system are recognized using the term 'Total Dynamic Head', TDH. The pump reacts to a change in the system. For example, in a small system, this could be the changes in tank levels, pressures, or resistances in the piping. In a large system, an example would be potable water pumps designed for an urban area consisting of 200 homes. If after 5 years the same urban area has 1,000 homes, then the characteristics of the system have changed. New added piping adds friction head (Hf). There could be new variations in the levels in holding tanks, affecting the static head (Hs). The increase in flow will affect the pressure head (Hp), and the increased flow in old, scaled piping will change the velocity head (Hv). New demands in the system will move the pumps on their curves. Because of this, we say that the system controls the pump. And if the system makes the pump do what it cannot do, then the pump becomes problematic, and will spend too much time in the shop with failed bearings and seals.

The elements of the Total Dynamic head (TDH)

The Total Dynamic Head (TDH) of each and every pumping system is composed of up to four heads or pressures. Not all systems contain all four heads. Some contain less than four. They are:

1. **Hs** – the static head, or the change in elevation of the liquid across the system. It is the difference in the liquid surface level at the suction source or vessel, subtracted from the liquid surface level where the pump deposits the liquid. The Hs is measured in feet of elevation change. Some systems do not have Hs or elevation

change. An example of this would be closed systems like water in the radiator of your car. Another example would be a swimming pool re-circulating filter pump. The vessel being drained (the pool) is the same level as the vessel being filled (the pool). If there is a difference in elevation across the system, this difference is recorded in feet and called Hs.

2. **Hp** – the pressure head, or the change in pressure across the system. It is expressed in feet of head. The Hp also may, or may not exist in every system. If there is no pressure change across the system, then forget about it. An example of this would be a recirculated closed loop. Another example would be if both the suction and discharge vessels have the same pressure. Think of a pump draining a vented atmospheric tank, and filling a vented atmospheric vessel. The atmospheric pressure would be the same on both vessels, thus no Hp. If Hp is present, then note the pressure change and employ it in the following formula. Sometimes, it is necessary to use a pump to drain a tank at one pressure (like atmospheric pressure), while filling a tank that might be closed and pressurized. Think of a boiler feed water pump where the pump takes boiler water from the deaerator (DA) tank at one pressure, and pumps into the boiler at a different pressure. This is a classic example of Hp. The formula is:

$$\mathbf{Hp} = \frac{\Delta \mathbf{psi} \times \mathbf{2.31}}{\mathbf{sp.\ gr.}}$$

where: Δpsi = boiler pressure – DA tank pressure

3. **Hv** – the velocity head, or the energy lost into the system due to the velocity of the liquid moving through the pipes. The formula is:

$$\mathbf{Hp} = \frac{V^2}{2g}$$

where: V = velocity of the fluid moving through the pipe measured in feet per second, and **g** = the acceleration of gravity, 32.16 ft/sec^2

AUTHOR'S NOTE

Hv is normally an insignificant figure, like a fraction of a foot of head or fraction of a psi, which can't be seen on a standard pressure gauge. But you can't forget about it because it is needed to calculate the friction head. If the Hv converts to a pressure that can be observed on a standard pressure gauge, like 6 or 10 psi, the problem is the inadequate pipe diameter.

4. **Hf** – friction head is the friction losses in the system expressed in feet of head. The Hf is the measure of the friction between the pumped liquid and the internal walls of the pipe, valves, connections and accessories in the suction and discharge piping. Because the Hv and the Hf are energies lost in the system, this energy would never reach the final point where it is needed. Therefore these heads must be calculated and added to the pump at the moment of design and specification. Also it's necessary to know these values, especially when they're significant, at the moment of analyzing a problem in the pump. The Hf and the Hv can be measured with pressure gauges in an existing system (see the Bachus & Custodio formula in this chapter). If the system is in planning and design stage and does not physically exist, the Hf and Hv can be estimated with pipe friction tables (ahead in this chapter). The Hf formula for pipe is:

$$\mathbf{Hf} = \frac{\mathbf{Kf \times L}}{\mathbf{100}}$$

where: Kf = friction constant for every 100 ft of pipe derived from tables **L** = actual length of pipe in the system measured in feet.

The Hf formula for valves and fittings is:

$$\mathbf{Hf = K \times Hv}$$

where: **K** = friction constant derived from tables, and $\mathbf{Hv} = \dfrac{\mathbf{V^2}}{\mathbf{2g}}$

The sum of these four heads is called the total dynamic head, TDH.

$$\mathbf{TDH = Hs + Hp + Hf + Hv}$$

The reason that we use the term 'dynamic' is because when the system and the pump is running, the elevations, pressures, velocities, and friction losses begin to change. In other words, they're dynamic.

AUTHOR'S NOTE

When the system is designed, the engineer tries to find a pump that's BEP is equal to or close to the system's TDH (the system's TDH ≅ BEP of the pump). But once the pump is started, the system becomes very dynamic, leaving the poor pump with a static BEP.

The purpose of the system curve is to graphically show the elements of the TDH imposed on the pump curve. The system curve shows the complete picture of the dynamic system. This permits the purchase, installation, and maintenance of the best pump for the system. The system curve is most useful when mated with the pump family curve. This is why the family curves are the most useful to the design engineer, the maintenance engineer, and purchasing personnel.

At the beginning of this chapter, we stated that the system governs the pump. This being the case, the pump always operates at the intersection of the system curve and the pump curve. And the goal of the engineer is to do everything possible to assure that this point of intersection coincides with the pump's BEP. Consider the following graph (Figure 8–1).

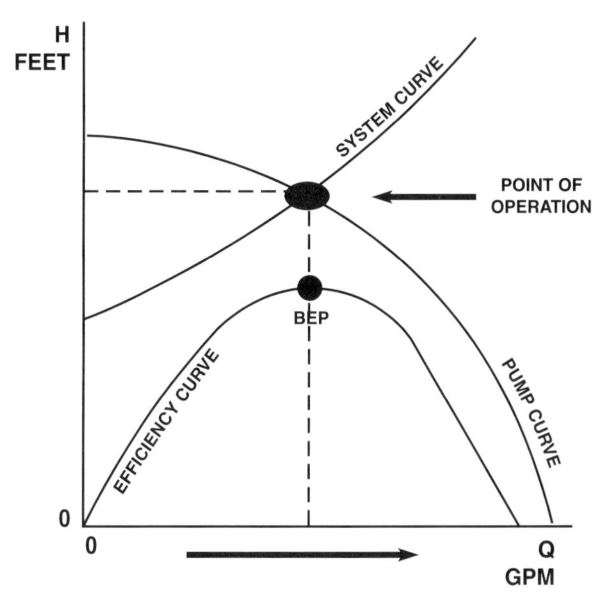

Figure 8–1

It is necessary to understand the TDH and it's components in order to make correct decisions when parts of the system are changed, replaced, or modified (valves, heat exchangers, elbows, pipe diameter, probes, filters, strainers, etc.) It's necessary to know these TDH values at the moment of specifying the new pump, or to analyze a problem with an existing pump. In order to have proper pump operation with low maintenance over the long haul, the BEP of the pump must be approximately equal to the TDH of the system.

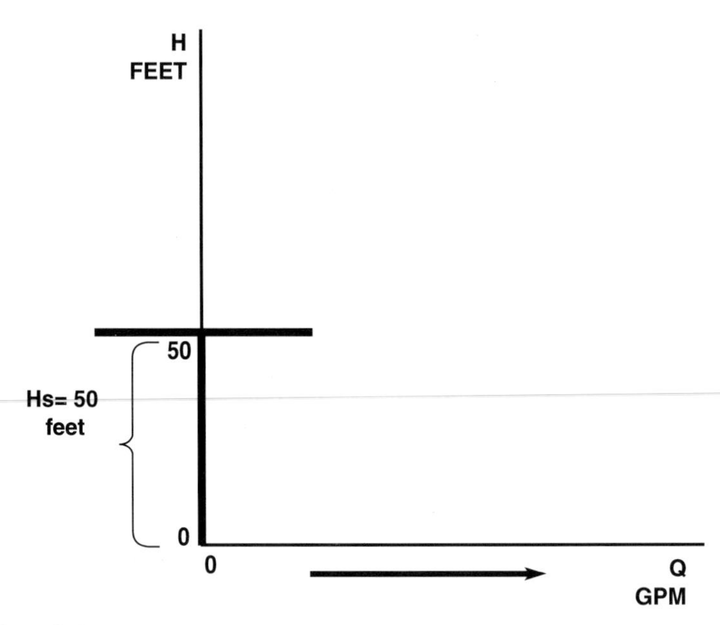

Figure 8–2

Determining the Hs

Of the four elements of the TDH, the Hs and the Hp (elevation and pressure) exist whether the pump is running or not. The Hf and the Hv (friction and velocity losses) can only exist when the pump is running. This being the case, we can show the Hs and the Hp on the vertical line of the system curve at 0 gpm flow. The Hs is represented as a T on the graph below. For example, if the pump has to elevate the liquid 50 feet, the Hs is seen in Figure 8–2.

Determining the Hp

The Hp also can exist with the pump running or off. We can represent this value with an O or oval on the vertical line of the below graph. The Hp is added to and stacked on top of the Hs. Let's say that our system is pumping cold water and requires 50 ft of elevation change and 10 psi of pressure change across the system. Now, our pump not only has to lift the liquid 50 ft, but it must also conquer 23 ft of Hp. Remember that 10 psi is 23.1 ft of Hp:

$$\mathbf{Hp} = \frac{\mathbf{10\ psi \times 2.31}}{\mathbf{sp.\ gr.}}$$

Here is the system curve showing Hs and Hp (Figure 8–3).

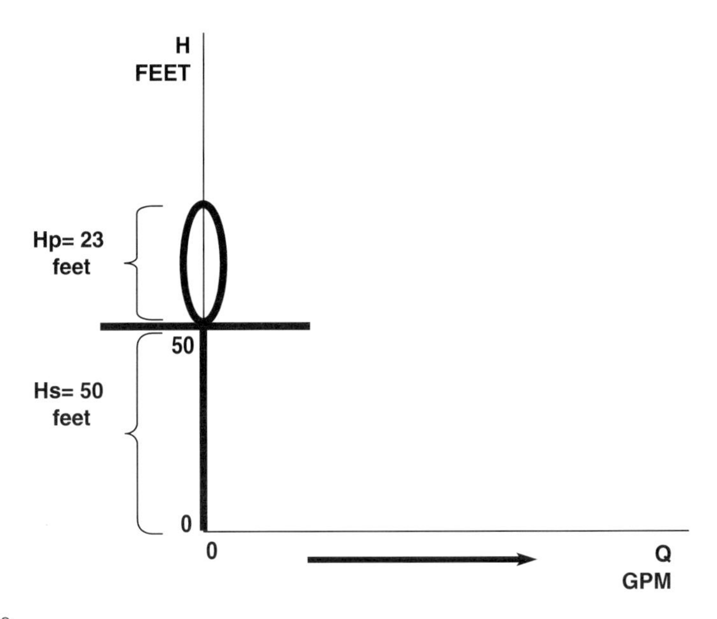

Figure 8–3

Calculating the Hf and Hv

Continuing with our example, before starting the system we already know that the pump must comply with 73 ft of static and pressure head. At the moment of starting the pump, the elements of Hf and Hv come into play as flow increases. Remember that Hf and Hv work in concert because the Hv is used to calculate the Hf. These values can be calculated using a variation on the Affinity Laws. The Affinity Laws state that the flow change is proportional to the speed change ($Q \alpha N$), and that the head change is proportional to the square of the speed change ($H \alpha N^2$). Therefore algebraically, the head change is proportional to the square of the flow change ($\Delta H \, \alpha \Delta Q^2$). Also, the friction head change and velocity head change are proportional to the square of the change in flow (ΔHf and $\Delta Hv \, \alpha \Delta Q^2$). On the system curve, the Hf and Hv begin at 0 gpm at the sum of Hs and Hp, and rise exponentially with the square in the change in flow. On the graph, it is seen as in Figure 8–4.

In a perfect and static world, we could apply the Affinity Laws to calculate the Hf and Hv, and calculate how the Hf and Hv change by the square of the change in flow. Well, the world is neither perfect, nor is it static. And, pipe is not uniform in its construction.

Some engineers (who normally are precise and specific) are charged with the task of approximating the friction losses (the Hf and Hv) in

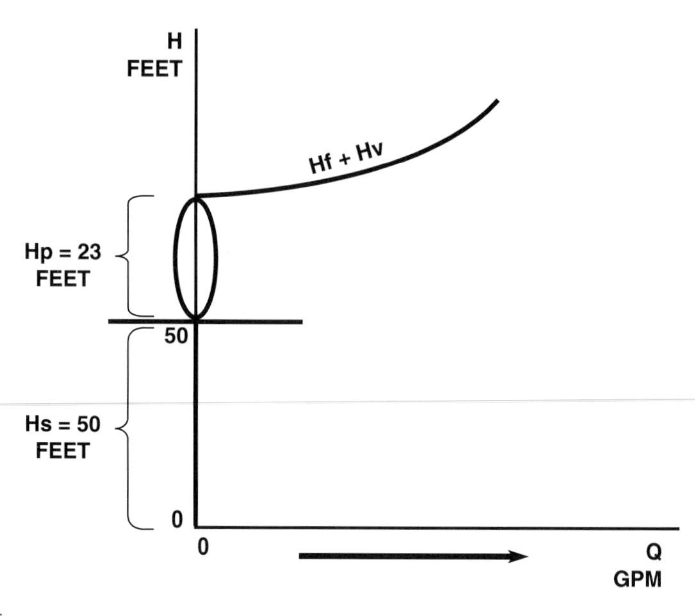

Figure 8–4

piping before the system exists. In the design stage, when the system exists only in drawings and plans, the civil engineer knows the proposed heads and elevations. And, he knows the proposed pressures in the system under construction. But he does not know, nor can he calculate the friction and velocity losses with the variations in pipe construction. Over the years, civil engineers have found refuge in the 'Hazen and Williams' Formula, and also the 'Darcy/Weisbach' Formulas for estimating the friction (Hf), and velocity (Hv) losses in proposed piping arrangements.

The Hazen and Williams formula

Mr. Hazen and Mr. Williams were two American civil engineers from New England in the early 1900s. In those days, piping used to carry municipal drinking water was ductile iron, coated on the inside diameter with tar and asphalt. The tar coating gave improved flow characteristics to the water compared to the flow characteristics of the ductile iron piping without the coating. The engineers Hazen and Williams derived their formula, a variation on the Affinity laws, and introduced a correction factor for friction losses of about 15%. Simply put, their formula is: $\Delta Hf \; \alpha \; \Delta Q^{1.85}$. The H & W method is the most popular among civil and design engineers. The formula is empirical, simple, and easy to apply. It is the method to calculate friction losses that is required by most of the municipal water agencies. The H & W formula assumes a turbulent flow of water at ambient temperature. As an approximation, it is most precise with velocities between 3 and 9 feet per second in pipes with diameters between 8 and 60 inches.

The Darcy/Weisbach Formula

This formula is another variation on the Affinity Laws. Monsieur's Darcy and Weisbach were hydraulic civil engineers in France in the mid 1850s (some 50 years before Mr. H & W). They based their formulas on friction losses of water moving in open canals. They applied other friction coefficients from some private experimentation, and developed their formulas for friction losses in closed aqueduct tubes. Through the years, their coefficients have evolved to incorporate the concepts of laminar and turbulent flow, variations in viscosity, temperature, and even piping with non uniform (rough) internal surface finishes. With so many variables and coefficients, the D/W formula only became practical and popular after the invention of the electronic calculator. The D/W formula is extensive and complicated, compared to the empirical estimations of Mr. H & W.

AUTHOR'S NOTE

The merits of the Hazen and Williams's formula versus the Darcy/Weisbach formula are discussed and argued interminably among civil engineers. It is our opinion that if a student learned one method from his university professor, normally that student will prefer to continue using that method. The two formulas are variations on the Affinity Laws, which are probably equally adequate to 'guestimate' the friction losses in non-uniform piping. Both the H & W and D/W formulas try to approximate the friction losses (Hf and Hv) in a piping system that physically does not exist. It doesn't exist because these calculations occur during the design phase of a new installation. But in this phase, it is necessary to begin specifying the pumps, although based on incomplete information. It's somewhat like a blind man throwing an invisible dart at a moving dartboard.

It really doesn't matter which formula (the H & W or the D/W) one prefers to use in calculating friction losses (Hf and Hv) in a pipe. Both formulas have deficiencies. Both formulas assume that all valves in the system are completely and totally open (and this is almost never the case). Both formulas assume that all instructions on construction and assembly (the pipes, supports, connections, valves, elbows, flanges and accessories) are followed to the letter (practically never). Both formulas assume that there are no substitutions during construction and assembly due to back orders and delivery shortages (Yeah, right!). Neither formula considers that scale forms inside the piping and that the interior diameters, thus Hf and Hv, will change over time. Neither formula considers that control valves are constantly manipulated, nor that filters clog. One formula doesn't consider that viscosity, thus stress and friction, can change with temperature or agitation. And both formulas are based on municipal water with piping adequate for that service only.

In recent years new equipment has been invented, chemical processes, piping materials, valve designs, and new technologies not considered when these formulas were developed with cold water in the 19th century. There is a need to measure the actual losses once an industrial plant is commissioned and operations begin. The authors of this book have developed a formula that permits the measurements of these losses in a live functioning system. Here it is:

The Bachus & Custodio Formula

Also called the DUH!! Factor. You'll need to gather information from pressure gauges mounted to the existing system. With the previously mentioned formulas, the Hf and the Hv are estimated in the initial phase when everything is new. The Bachus & Custodio method measures the exact Hf and Hv in any existing system. It doesn't matter when it was built.

The **Bachus & Custodio** Formula is the following:

$$\textbf{System Hf and Hv} = \left(\frac{(\Delta PDr - \Delta PDo) + (\Delta PSr - \Delta PSo) \times 2.31}{sp.\ gr.}\right)$$

where: **ΔP** = pressure differential from an upstream to a downstream gauge in a section of pipe **Dr** = Discharge Running, the discharge piping with the pump running. **Do** = Discharge Off, the discharge piping with the pump not running. **Sr** = Suction Running, the suction piping with the pump while running. **So** = Suction Off, the suction piping with the pump not running. **2.31** = conversion factor between psi and feet of elevation **sp.gr.** = Specific Gravity

In general, the Hf and Hv are observed while considering the system. We'll see this further ahead. The Hf and the Hv are the reasons that companies contract civil engineers to design their new plants. Years later, those design parameters have changed due to erosion and other factors. Let's look at the following situation, pumping a liquid from one tank into another.

This system, although simple, with only one pump, is more or less representative of all systems. This system is composed of 180 ft of pipe; 40 ft of 6 inch suction pipe, and 140 ft of 4 inch discharge pipe.

This system piping uses fittings with bolted flanges, see Figure 8–5. The 6 inch elbows have a constant (K value) of 0.280. The 4 inch elbows have a K value of 0.310. The 6 inch gate valves have a K value of 0.09. The 4 inch gate valves have a K value of 0.15. The 4 inch globe valve

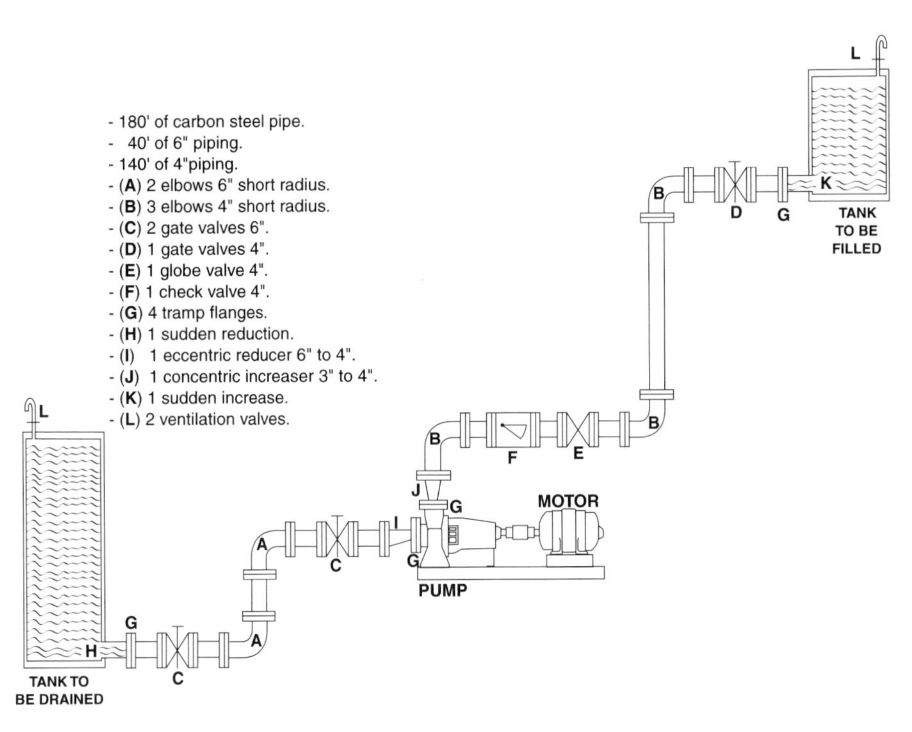

Figure 8–5

has a K value of 6.4. The 4 inch check valve has a K value of 2.0. The 4 inch tramp flanges have a K value of 0.033. The 3 inch tramp flange has a K value of 0.04. The sudden reduction has a K value of 0.5. The 6 to 4 inch eccentric reducer has a K value of 0.28. The 3 to 4 inch concentric increaser has a K value of 0.192. The sudden increase has a K value of 1.0. The required flow is 300 gallons per minute. The constants mentioned (K) are given values provided by manufacturers and can be found on charts provided by different organizations.

The goal is to apply the formulas, the K values, and the pipe and connections friction values to determine the Hf and Hv, plus the Hs and Hp, and then the TDH, total dynamic head in the system. Then we can specify a pump for this application.

One component of the TDH is the Hs, the static head. In this example the surface level in the discharge tank is 115.5 ft above the pump centerline. The surface level in the suction tank is 35.5 ft above the pump centerline. The ΔHs, by observation is 80 ft. See Figure 8–6.

Another component of the TDH is the Hp, pressure head. We can see in Figure 8–7 that both tanks have vent valves. These two vessels are exposed to atmospheric pressure, which is the same in both tanks. So by simple observation, pressure head doesn't exist. ΔHp = 0.

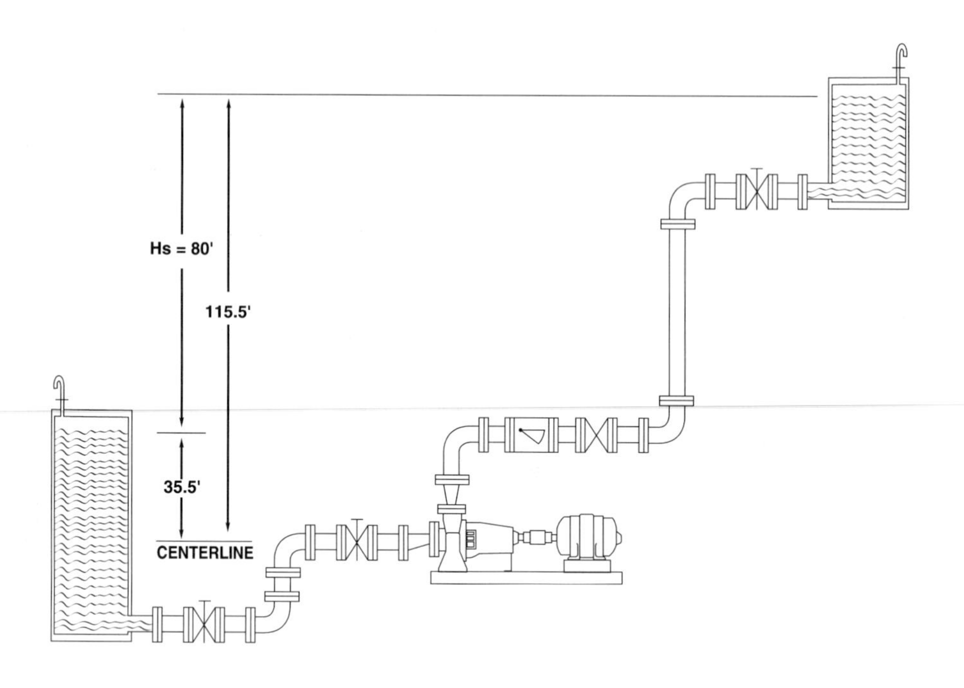

Figure 8-6

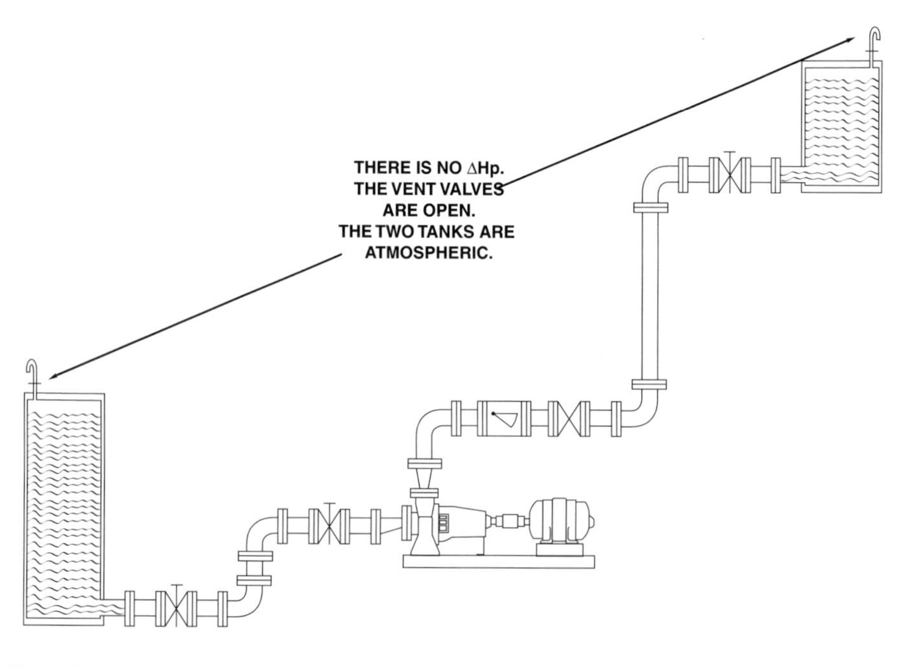

Figure 8-7

AUTHOR'S NOTE

The following is not very entertaining to read. The authors have included this section so that the readers can gain an appreciation for the detailed work of the design engineer on calculating the frictions and velocities in a piping system. Admittedly, there are computer programs today that will perform these calculations in a flash. But 30 years ago and before, these calculations were done with mental software (the engineer's brain), a mechanical computer (a slide rule), and a manual printer (pencil and paper).

In most cases, the design engineer and architect begin with an open field of rabbits and weeds. Two years later there is a hotel, gasoline station, or paint factory built on the site of the open field. And the day that the new owners open their hotel, or start mixing paints, most of the new pumps are running within 5% of their best efficiency points. The pumps were mostly designed correctly into their new systems, and run for various years without problems, an amazing feat of engineering, math, and art ... before computers.

Remember that we're calculating the TDH. Two elements of the TDH, the Hs and the Hp, were determined mostly by observation of the system and drawings. The remaining two elements, the Hf and the Hv, are the most illusive and difficult to calculate. Yet, they determine how and where the pump will operate on its curve. Continue reading.

Using the formulas, the K values, and the pipe schedule tables found in the Hydraulic Institute Manual, ($V_{suction}$ = 3.33 ft/sec for 6 inch pipe @ 300 GPM and $V_{discharge}$ = 7.56 ft/sec for 4 inch pipe @ 300 GPM) or other source, we can estimate or calculate the friction and velocity heads in the system. Because the Hv is used to calculate the Hf, we'll begin with the Hv. The formula is:

System Hv = Hv suction + Hv discharge

= $V^2 \div 2g$ suction + $V^2 \div 2g$ discharge

= $3.33^2 \div 64.32 + 7.56^2 \div 64.32$

= 0.172 ft. suction + 0.888 ft. discharge

System Hv = 1.06 feet

The **Hv suction** and **discharge** values will be used in the **Hf** formula.

System Hf = Hf pipe + Hf elbows + Hf valves + Hf tramp flanges + Hf other

Taking this formula in groups, we begin with the Hf pipes.

Hf system piping = Hf suction piping+ Hf discharge piping.

= (K suction × L) ÷100 + (K discharge × L) ÷ 100

= (4.89 × 40) ÷100 + (.637 × 140) ÷100

= 1.956 + 0.891

Hf system piping = 2.848 feet

Now we calculate the Hf in the elbows

The formula is:

Hf elbows = Hf suction elbows + Hf discharge elbows

= 2 × 0.280 × 0.172 + 3 × 0.310 × 0.888

= 0.096 + 0.82

Hf elbows = 0.916 feet

Next, we calculate the Hf for the valves

There are 5 valves in all. There are two 6 inch gate valves in the suction pipe. There is a 4 inch gate valve, a 4 inch globe valve, and a 4 inch check valve in the discharge pipe. The formula is:

Hf system valves = Hf suction valves + Hf discharge valves

$$= K_{6''\ gate} \times Hv_{suction} + K_{4''\ gate} \times Hv_{disch.} + K_{4''\ check} \times Hv_{disch.} + K_{4''\ globe} \times Hv_{disch.}$$

= (2 × .09 × 0.172) +(1 × 0.16 × 0.888) + (1 × 2 × 0.888) + (1 × 6.4 × 0.888)

= 0.031 + 0.142 + 1.776 + 5.683

Hf system valves = 7.632 feet

Next we calculate the Hf in the tramp flanges in the system

A tramp flange is an unassociated flange or union. In the friction tables, valves, elbows, and other fittings are categorized as to whether they are flanged or screwed. This means they connect to the piping either by a bolted flange, or screwed into the pipe with male and female threading. For example, the friction losses through a 2 inch flanged elbow, or a 4 inch check valve, already takes into account the losses at the entrance and exit port fittings. Then there are unassociated 'tramp' flanges and unions. Examples would be unions between two lengths of pipe, or between a pipe and a tank, or between a pipe and a pump. They must be calculated because there is friction (and energy lost) as the fluid passes through a union. In our simple system, there is a 6 inch tramp flange on the suction pipe with the tank, and a 4 inch tramp with the pump. There's a 3 inch tramp flange at the pump discharge and another 4 inch tramp at the discharge tank. The formula is:

Hf system tramp flanges = **Hf suction tramps + Hf discharge tramps**

= $K_{6''} \times Hv_{suct.} + K_{4''} \times Hv_{suct.} + K_{4''} \times Hv_{disch.} + K_{3''} \times Hv_{disch.}$

= $(0) + (0.033 \times 0.172) + (0.033 \times 0.888) + (0.04 \times 0.888)$

= $0 + 0.005 + 0.029 + 0.035$

Hf system tramp flanges = **0.007 feet**

Admittedly, Hf of 0.007 foot is an insignificant number. Think of it this way. With only one pump and less than 200 ft of pipe in our simple system, there are four tramp unions. Imagine an oil refinery with 20,000 pumps and thousands of miles of pipe and equipment on site. Imagine the number of tramp flanges in the fire water system in a skyscraper building. In a real set of circumstances the Hf values through tramp flnages unions could be significant, and they would have to be calculated to specify the correct pumps.

Last, we need to calculate the Hf losses through other connections in the piping

There is a sudden reduction in the suction between the tank and the piping. There is an eccentric 6-to-4 reducer between the suction pipe and the pump. There is a concentric 3-to-4 increaser from the pump back into the piping, and a sudden enlargement going into the discharge tank. The formula is:

Other Hf = $Hf_{\text{sudden reduction}} + Hf_{\text{eccentric reducer}} + Hf_{\text{concentric increaser}} + Hf_{\text{sudden enlargement}}$

= $(0.05 \times 0.172) + (0.28 \times 0.172) + (0.192 \times 0.888) + (1 \times 0.888)$

= $0.086 + 0.048 + 0.170 + 0.888$

Other Hf = **1.192 feet**

Now we have all the information to calculate the Hf in the system and then the TDH of the system. Once again:

System Hf = **Hf pipe + Hf elbows + Hf valves + Hf flanges + Hf other**

= **2.848 ft + 0.916 ft + 7.632 ft + 0.007 ft + 1.192 ft**

= **12.595 ft**

Consider all the mathematical gyrations required just to determine the Hv and Hf. This is a lot of math for one pump. Imagine the work to specify pumps for a paper mill or beer brewery or municipal water system. Now you can see why governments and pharmaceutical companies contract consulting engineering companies to do this work and specify the pumps. Finally, we can calculate the **TDH** of the system:

TDH = Hs + Hp + Hf + Hv

= 80 ft + 0 ft + 12.595 ft + 1.06 ft

= 93.655 ft

This system requires a pump with a best efficiency point (BEP) of 94 feet at 300 gallons per minute. If this is a conventional industrial centrifugal pump with a BEP of 94 feet, the shut-off head should be approximately 110 feet. And if the motor is a standard NEMA four-pole motor spinning at about 1800 rpm, the diameter of the impeller should be approximately 10.5 inches. If this pump were bought off the shelf from local distributor stock, it would probably be a 3 × 4 × 12 model end-suction centrifugal back pullout pump with the impeller machined to about 10.5 inches before installing the pump into the system. And that's the way it is done.

If the system already exists and the equipment is running, we can recover the **Hf** and **Hv** from gauges using the Bachus & Custodio Method, and forget about all those calculations. See Figure 8–8 opposite, with the corresponding elevations and placement of pressure gauges installed into the piping numbered 1 through 5.

In this system drawing, pressure gauges 1, 2, and 3 are in the suction piping. Gauges 4 and 5 are in the discharge piping. With the system and pump turned off, we would open the vent valves on both the suction and discharge tanks, this assures that both sides of the system are atmospheric and cancels the **Hp**. The discharge tank and all piping should be full with water for the test, or if required, the pumped liquid. Remember that gauge readings will be adjusted by the specific gravity. Expel all air bubbles in the piping. Some pumps have a little petcock valve to allow expelling any trapped air in the volute. On the pump, conventional stuffing boxes can also trap air. This must be expelled too. Vertical valve stems in the piping can trap air. Loosen the packing to expel this trapped air. This is done so that there is a complete column of liquid from the top to the bottom of the system. Air pockets and bubbles might cause inaccurate pressure gauge readings. All valves in the column (including the check valve) should be opened, except for the gate valve between gauges 1 and 2. It should be closed to hold the column of liquid and prevent draining the line.

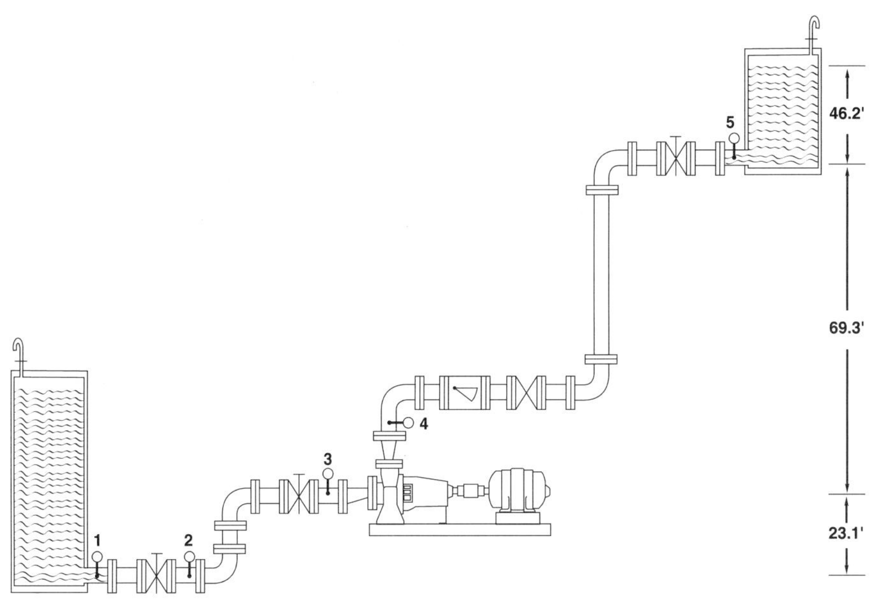

Figure 8-8

Here's a quick review of the Bachus & Custodio Formula:

System Hf and Hv = [(ΔPDr – ΔPDo) + (ΔPSr – ΔPSo) × 2.31] ÷ sp. gr.

Let's take our readings with water as the test liquid just to keep the conversions simple. With the system and pump off, note that gauge 5 should be reading 20 psi. This is because it is 46.2 feet below the surface level in the discharge tank. Confirm that gauge 4 is reading 50 psi. It is 115.5 feet deep into the column. The difference between gauges 4 and 5 is 30 psi. The ΔPDo = 30 psi.

In the suction line, note that gauge 3 is also reading 50 psi. It also has 115.5 feet of liquid elevation on it. Pressure gauge 2 should read 60 psi because it is 138.6 feet deep into the column. This indicates that the ΔPSo = 10 psi.

Gauge 1, on the other side of the closed valve, is reading the elevation in the suction tank. This gauge should be reading 25 psi because it is 58.6 feet deep into its column.

Now, open the gate valve between gauges 1 and 2. Start the pump motor, and relieve the check valve if it is being mechanically held open. Permit the pump to run a few minutes to stabilize, relieving any surging. We'll continue to note pressure gauge readings with the system functioning.

Because all valves are now open, gauge 1 becomes our upstream gauge on the suction line. With the pump running all activity on the suction side of the pump is separated from the activity on the discharge side of the pump. Gauge 1 continues to read 25 psi. Gauge 2 should also record 25 psi. Gauge 3 should now be reading 15 psi, because this gauge is 23.1 feet above gauges 1 and 2.

However, gauge 3 is recording 13 psi (it should be reading 15 psi) with the system running. The ΔPSr is 12 psi.

AUTHOR'S NOTE

Gauges 1 and 2 should be reading the same pressure with the system running, as gauge 1 was reading with the system off. If you're using precision digital pressure instrumentation gauges, gauge 2 might possibly record a fraction psi less. This is because gauge 2 is now recording minute losses between the tank and the gauge including losses through the opening into the pipe and the losses through the gate valve.

If there should be a divergence in the readings of the two gauges, something is out of control. There might be an obstruction at the tank drain line or maybe the gate valve is not totally open. Maybe the level has dropped in the tank. Maybe the vent valve on the tank top is not open. Maybe the gauges need calibration. Send them to a calibration shop a couple of times per year. But, isn't it interesting how much more you know about your system after learning to interpret the pressure gauges. Who is responsible for specifying, selling, and buying pumps without adequate instrumentation?

Now we consider the pump. We've already discussed in this book that the pump takes the energy that the suction gives it, the pump adds more energy, jacking the energy up to discharge pressure. In this case the pump is designed with a BEP of 94 feet, which also is the TDH of the system. The 94 feet indicate that the pump can generate about 40 psi at 300 gpm (94 ÷ 2.31 = 40.6 psi if the liquid is water). This is confirmed with a flow meter and a pump curve. The suction pressure is 13 psi. The discharge pressure gauge (4 gauge) should be reading 53 psi (40 + 13).

AUTHOR'S NOTE

The pump's discharge pressure is a function of the suction pressure. Regrettably, most pumps in the world don't have a gauge reading suction pressure. In our example here, if our pump is generating less than 40 psi, the pump is operating to the right of its BEP, and is losing efficiency. Was the pump assembled correctly? Was it repaired correctly, with all parts machined to their correct tolerances? Is the motor's velocity correct? Is there a flow meter installed? The pump is always on its curve. If this pump were generating more than 40 or 41 psi, it would be operating to the left of its BEP. Verify the other factors.

With gauge 4 on the pump discharge reading 53 psi, the 5 gauge should be reading 30 psi less, or 23 psi. This is because the 5 gauge is 69.3 feet above gauge 4. However gauge 5, by observation, is only reading 18 psi. Therefore ΔPDr = 35psi (53 – 18). We have all the information we need to insert into the Bachus & Custodio Formula:

$$\text{Hf \& Hv} = (\Delta PDr - \Delta PDo) + (\Delta PSr - \Delta PSo) \times 2.31 \div \text{sp. gr.}$$

$$= (35 - 30) + (12 - 10) \times 2.31 \div \text{sp. gr.}$$

$$= 5 + 2 \times 2.31 \div \text{sp.gr.}$$

$$= 7 \times 2.31 \div 1.0$$

$$\text{Hf \& Hv} = 16.17 \text{ feet}$$

The Bachus & Custodio Formula does not make mistakes. It is not based on models, or experiments developed 150 years ago. It doesn't depend on valves being completely open. It doesn't depend on the specific instructions regarding equipment assembly. It doesn't depend on new piping. It is not based on municipal water. It depends on the actual piping and other system fittings, as they are now, and the next shift, and tomorrow, and next month. If a resistance load changes, it will be registered on the gauges. If the pipe diameter changes, it is recorded on the gauges. If new equipment is added, it is visible on the gauges. The pressure gauges and other instrumentation are the pump's control panel. You wouldn't drive a car without a dashboard. Who is responsible for specifying, selling and buying pumps without adequate instrumentation? Regarding pump failure, problematic seals and bearings that need emergency maintenance: in about 80% of all cases, the pump is telling the operators what the problem is, hours, days and weeks before the failure event occurs. What's really happening is that no one is interpreting the information on the gauges.

Regarding the TDH, isn't it interesting that the Hs and the Hp are determined by simple observation? This detailed discussion on the Hf and Hv probably has the reader ready to throw this book into the garbage. With the Bachus & Custodio Formula, the differential pressure gauge readings on the system with the pump turned off, will cancel any elevation changes (Hs) existing in the system. Exposing both sides of the system to atmospheric pressure cancels the pressure changes (Hp). And then with the system operating and the pump turned on, the further differential gauge readings will record the Hf and Hv that are being lost in the system. Remember too, that the other mentioned resistance approximations, Hazen & Williams, and Darcy/Weisbach, are only valid in the first few hours or days of service. The system begins to change once the pump is turned on and production begins. Operators open and close valves to meter the flow through the pipes. Filters and strainers begin to clog. Inside pipe diameters form scale.

New equipment is installed. Other changes occur with maintenance. The equipment loses its efficiency. Install gauges on your pumps and teach the operators and maintenance personnel to interpret the information.

The dynamic system

Let's continue with system curves. Up to this point, all elevations, temperatures, pressures and resistances in the drawings and graphs of systems and tanks have been static. This is not reality. Let's now consider the dynamic system curve and how it coordinates with the pump curve.

Variable elevations

In the next graph we observe that at the beginning of the operation, the lower tank is full, and the work of the pump is to complete the distance between the surface level in the lower tank and the discharge elevation above at the upper tank. At the end of the operation, the lower tank is empty and the work of the pump is to complete the new distance between the two elevations. Consider the next graphic (Figure 8–9).

At the beginning of the operation, the work of the pump is to complete elevation Hs_1. This elevation becomes Hs_2 at the end of the operation.

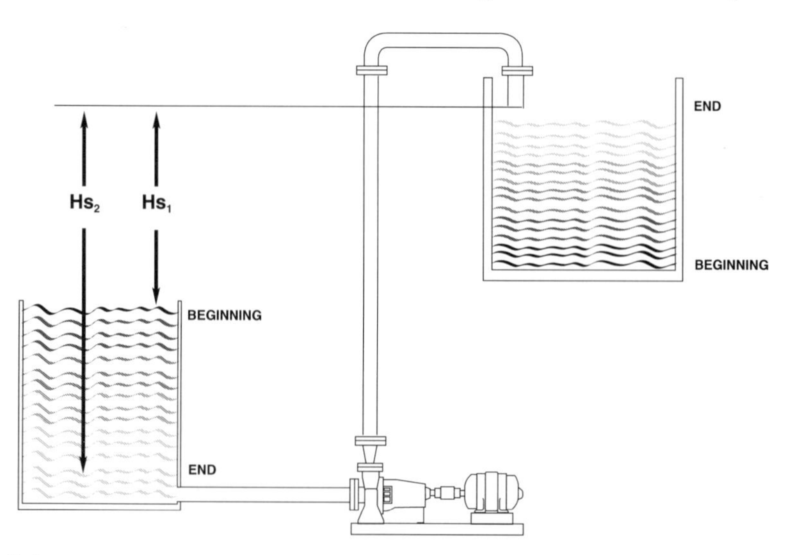

Figure 8–9

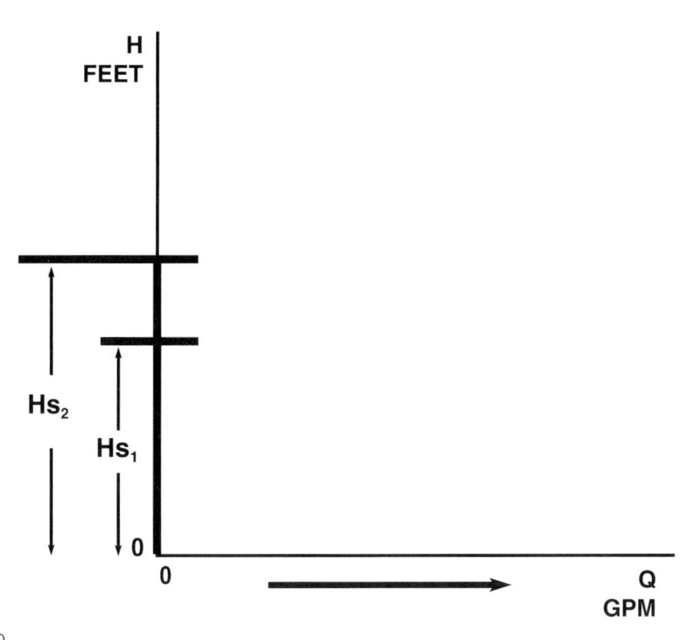

Figure 8–10

On starting the pump, and initiating flow we add the resistances in the pipes and they are shown above (Figure 8–10).

Now, due to the fact that both tanks are exposed to atmospheric pressure, there is no ΔHp to consider. Upon initiating flow, the Hf and Hv come into play and we have (Figure 8–11):

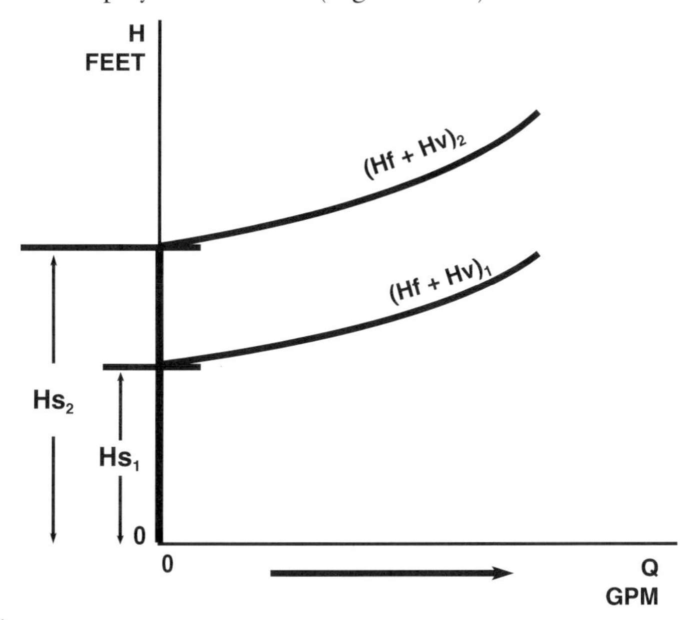

Figure 8–11

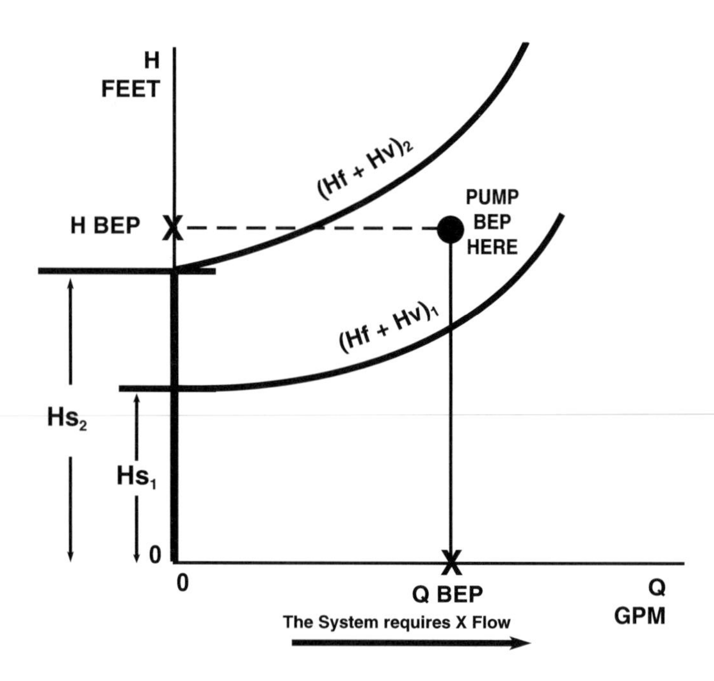

Figure 8–12

Next, we should find a pump who's BEP coordinates fall right between the Hs_1 and Hs_2 at flow X, as seen on the graph above (Figure 8–12).

With this information, the pump curve, coordinating with this system's demands according to the two tank levels, is seen this way (Figure 8–13).

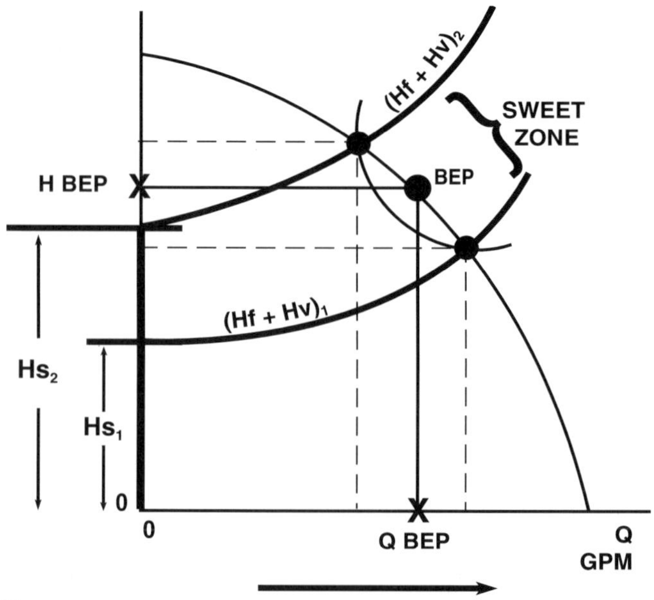

Figure 8–13

The happy zone

Now we can see the importance of the concentric ellipses of efficiency on the pump family curve. As much as possible we should find a pump whose primary efficiency arc covers the needs of the system. Certainly the needs of the system should fall within the second or third efficiency arcs around the pump's BEP. If the system's needs require the pump to consistently run too far to the left or right extremes on its curve, it may be best to consider pumps in parallel, or series, or a combination of the two, or some other arrangement, possibly a PD pump. We'll see this later.

As elevations change in the process of draining one tank and filling another, the pump moves on its curve from one elevation extreme to the other. If we've selected the right pump for the system, it will move from one extreme of its happy zone, through the BEP to the other extreme.

AUTHOR'S NOTE

This is the beginning of many problems with pumps. A pump is specified with the BEP at one set of system coordinates. Then the system (the TDH) goes dynamic, changing, and the pump moves on its curve away from its BEP out to one or the other extreme. It is necessary to determine the maximum and minimum elevations in the system and design the pump within these elevations. If the system continues to change on the pump, you'll either have to modify the system or modify or change the pump, unless you really like to change bearings and seals.

Dynamic pressures

Let's consider now a system with dynamic pressures and a constant elevation. A classic example of this would be where a pump feeds a sealed reactor vessel, or boiler. The fluid level in the reactor would be more or less static in relation to the pump. The resistances in the piping, the Hf and Hv, would be mostly static although they would go up with flow. The Hp, pressure head would change with temperature. Consider Figure 8–14.

The system curve, once again, is the visual graph of the four elements of the TDH. The Hp is stacked on top of the Hs. The Hp changes with a change in temperature in the reactor. If the reactor were cold, the Hp would be minimum or zero. We'll call this Hp_1. When the tank and fluid are heated, the Hp rises to its maximum. This is represented as Hp_2 on the graph (Figure 8–15).

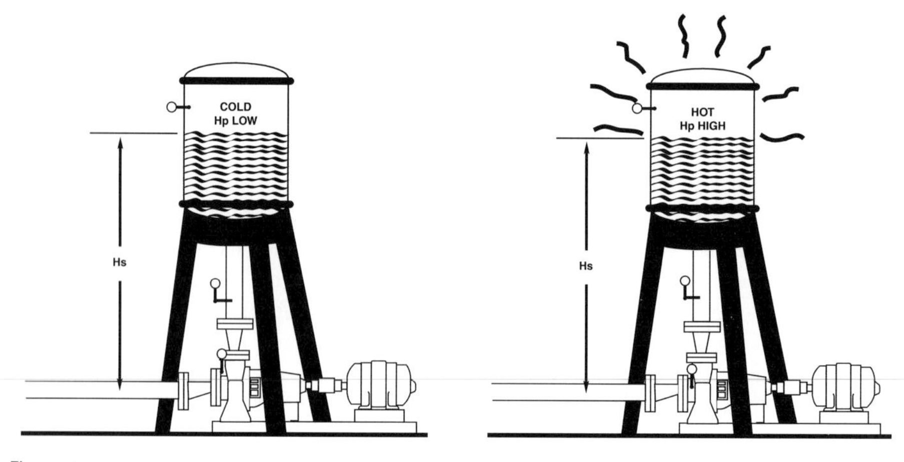

Figure 8–14

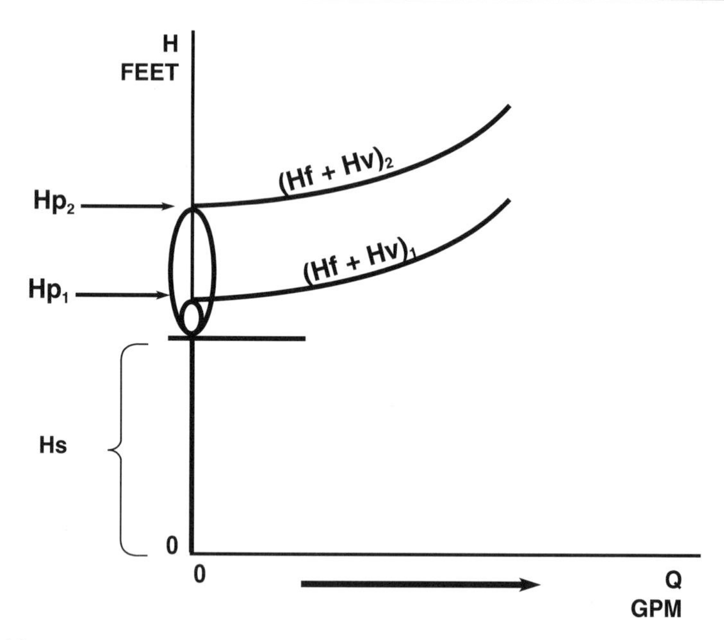

Figure 8–15

Let's say that the needs of the system require X flow. Now we search for a pump with a BEP at X gpm, at a head falling right between Hp_1 and Hp_2 on the system curve. See the next graph (Figure 8–16).

The system's ΔHp should fall within the pump's primary or secondary sweet zone. At the beginning of the operation, with the cold reactor vessel, the pump operates to the right of the BEP but within the sweet zone, and as the reactor vessel is heated, the pump migrates on its curve toward the left, crossing the BEP, to the other extreme of its sweet zone. When the reaction is completed and the tank cools, the pump

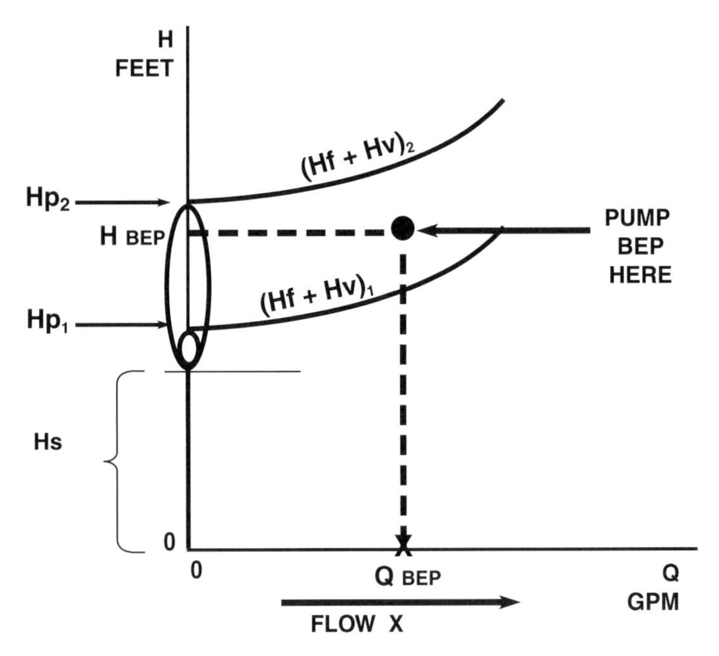

Figure 8–16

migrates again on its curve, this time toward the right, crossing the BEP and comes to rest on the right end of its sweet zone. See the next graph (Figure 8–17).

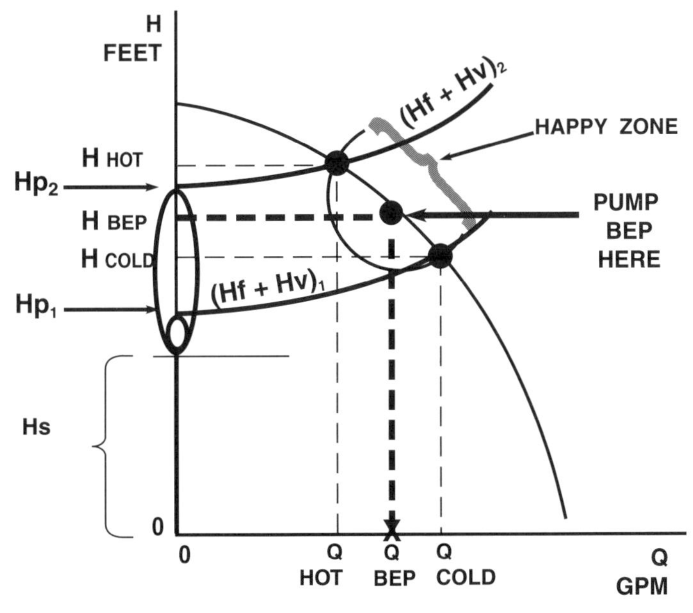

Figure 8–17

Again, we see the importance of the pump family curve, with its concentric ellipses of efficiency. It shows that in the beginning of the operation, the pump is operating to the right of its BEP. As the pressures rise in the system, the pump moves toward the left of the BEP. When the temperatures and pressures are reduced at the end of the process, the pump migrates again on its curve to the right of the BEP. During the entire operation, the pump is inside its primary or secondary sweet zone.

Variable resistances

The Hf and Hv represent the resistance losses in the system. Specifically, the Hf represents the energy consumed or lost due to friction in the system, and the Hv represents the energy consumed or lost due to the velocity of the fluid moving through the system. The Hf and the Hv are linked or connected because the Hv is used to calculate the Hf. If there is no velocity in the fluid, then the fluid is not moving through the pipes, and if the fluid is not moving through the pipes there can be no friction between the fluid and the internal pipe walls. If the resistance rises in the system, as in the case of a filter whose function is to clog over time, the flow is reduced through the filter, and the pressure or resistance rises. This means that the pump is moving toward the left on its curve. The resistances can change in the short term, or in the long term, with operations, with maintenance, or with a design change. Let's see how:

Short term resistance changes

The resistance in the pipes and system can change suddenly or in the short term due to a design change, operation, or maintenance. For example, many systems are designed incorporating variable speed motors, VFDs, to control production in a plant. The resistance is multiplied 4 times simply by doubling the velocity of the fluid in a pipe. Sometimes, in an existing system, the engineer orders to install a new control. For example, installing a flow meter into a pipe increases the resistance and the pump moves on its curve. In an effort to improve the final product, a production engineer orders a change to the screen meshes in the filters. This changes the Hf and the Hv in the system and the pump migrates on its curve.

In some plants, the operators have free reins to govern the flow in the pipes by opening and closing flow control valves. Strangling a valve reduces the flow and increases the resistance and pressure in the system in front of the valve. The pump moves away from the sweet zone of efficiency.

In a maintenance function, working against the production clock, someone changes a globe valve for a gate valve. A globe valve has between 20 and 40 times more resistance than a gate valve. Someone orders to exchange a bolted flanged long radius pipe elbow, with a welded mitered 90° elbow. This affects the resistance in the system and the pump on its curve.

AUTHOR'S NOTE The authors recommend that all plant personnel including the engineers, operators, and mechanics receive training to recognize these rapid, unexpected, quick changes in a system. Some of these changes can be controlled within certain limits. Others must be avoided as standard plant procedure. Almost no one in maintenance or reliability relates today's failed mechanical seal with the inoffensive change in a pipe diameter six weeks ago. Engineers should train their personnel to understand the result of these inadvertent changes. These rapid changes in a system are the source of pump maintenance.

Long term resistance changes

In the long term, filters and strainers become clogged: this is their purpose. Minerals and scale start forming on the internal pipe walls and this reduces the interior diameters on the pipe. A 4 inch pipe will eventually become a 3.5 inch pipe. This moves the pump on its curve because as the pipe diameter reduces, the velocity must increase to maintain flow through a smaller orifice. The Hf and Hv increase by the square of the velocity increase.

Also in the long term, the equipment loses its efficiency, and replacement parts are substituted in a maintenance function. Also, the plant goes through production expansions and contractions: new equipment is added into the pipes. In short, the system and its elevations and pressures, its resistances and velocities, are very dynamic. The BEP of the pump is static.

What must be done is establish the maximum flow, and the minimum flow, and implement controls. Regarding filters, you've got to establish the flow and pressure (resistance) that corresponds to the new, clean filter, and determine the flow and resistance that represents the dirty filter and its moment for replacement. These points must be predetermined. The visual graph of the system curve with its dynamic resistances are seen in this example filtering and recirculating a liquid in a tank. Consider the following graphs (Figures 8–18 and 8–19).

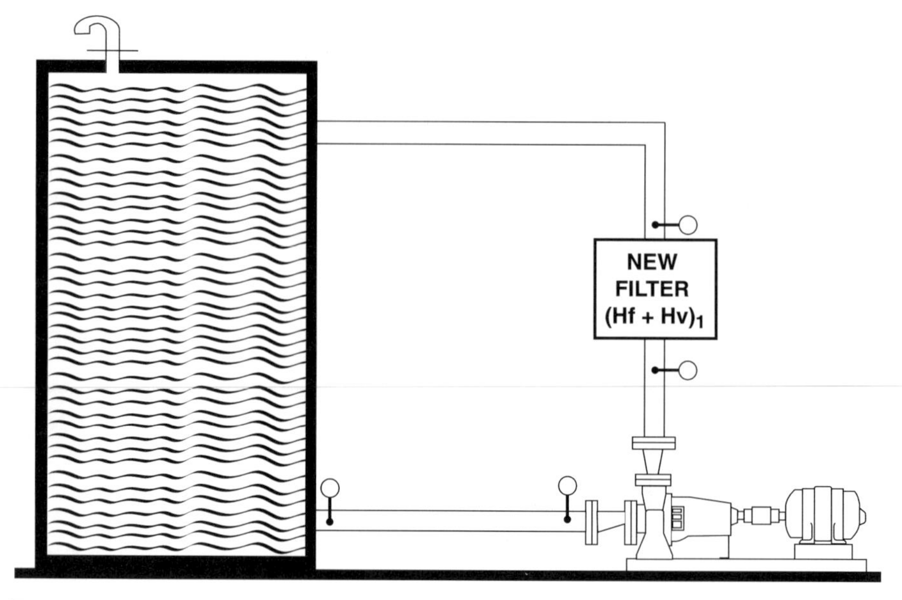

Figure 8–18

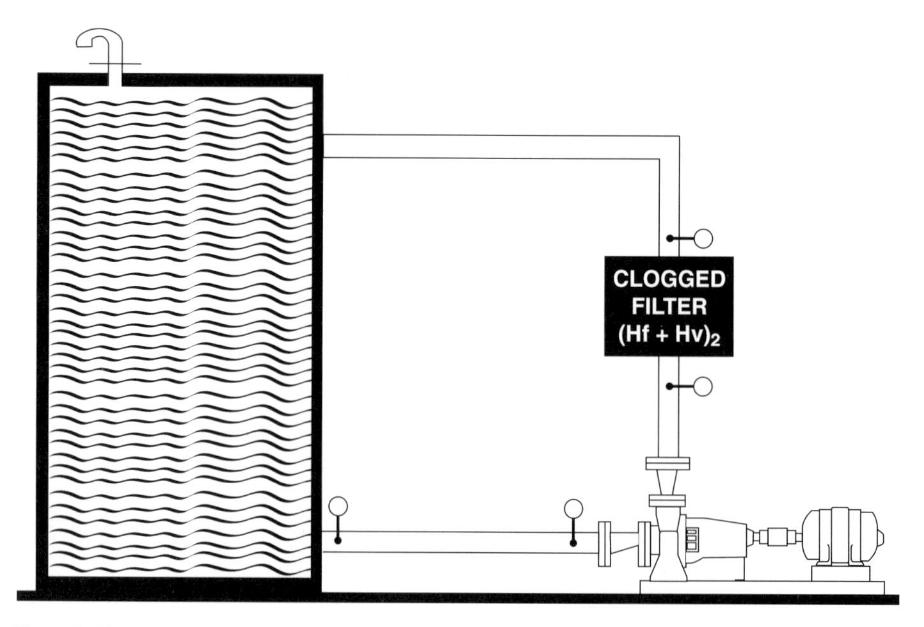

Figure 8–19

As mentioned earlier, the system curve with the clean and dirty filters should coincide within the sweet zone of the pump on its curve. (Figure 8–20 and Figure 8–21).

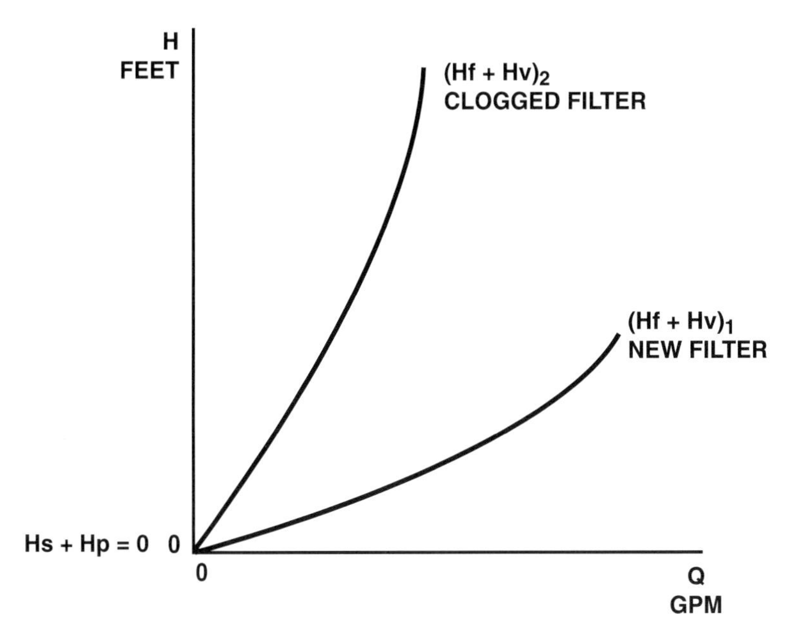

Figure 8–20

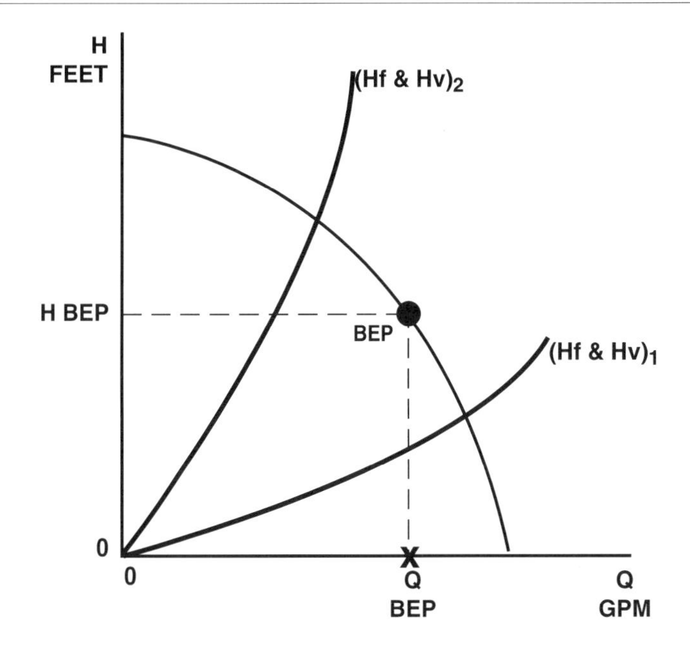

Figure 8–21

The pump will run to the right of its BEP within its sweet zone with the new filter, and slowly over time, move toward the left crossing the BEP as the filter screen clogs (Figure 8–21 and Figure 8–22).

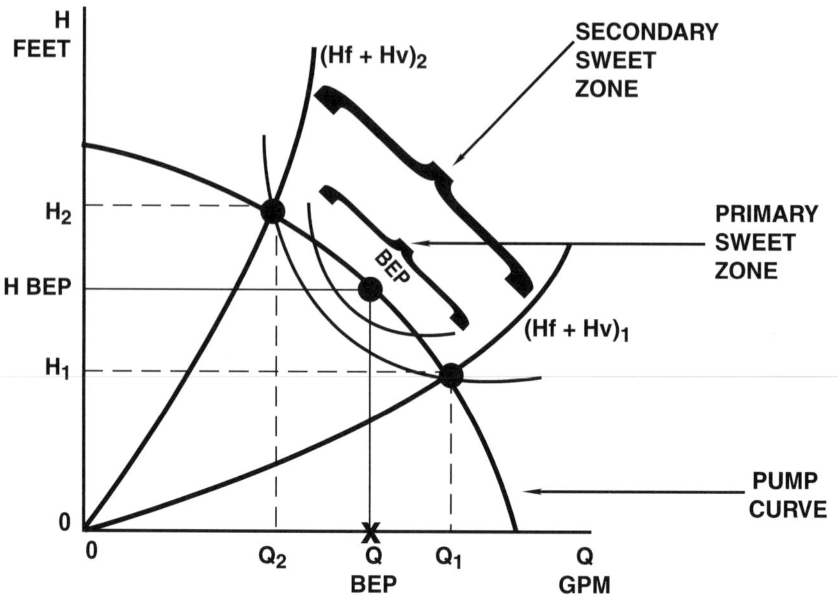

Figure 8–22

On superimposing the curve of a single pump over this system curve, we see that the system extremes are too wide for the pump to cover on its curve (Figure 8–22).

You should install pressure sensors that transmit a message to shut-off the pump, sound an alarm, or indicate to the operator that the moment to change the filter has arrived. With a new filter installed, the pump begins operating again to the right of the BEP within the sweet zone and slowly over time proceeds moving toward the other end of the sweet zone.

Pumps in parallel and pumps in series

Up to this point we've considered dynamic elements in the system with other elements static. There are times, and systems where everything is moving in concert together, with elevations rising and falling, variable pressures, clogging filters, and control valves opening and closing. When the entire system is dynamic, you've got to determine the elevation extremes, the pressure extremes, and the resistance extremes. The totally and completely dynamic system appears as Figure 8–23 and Figure 8–24.

When this happens, you need to consider an arrangement of pumps running in parallel, or in series, or in a combination of the two. Pumps in parallel are two or more pumps working side by side, taking the

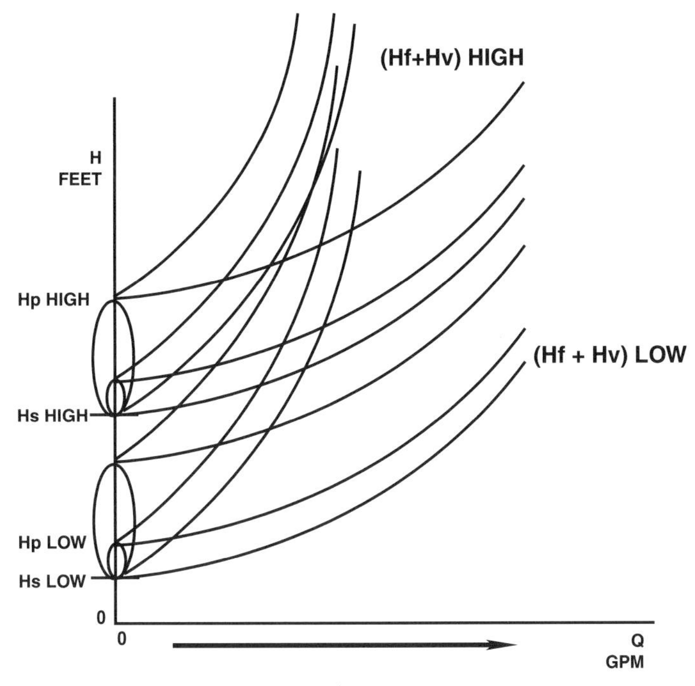

Figure 8–23

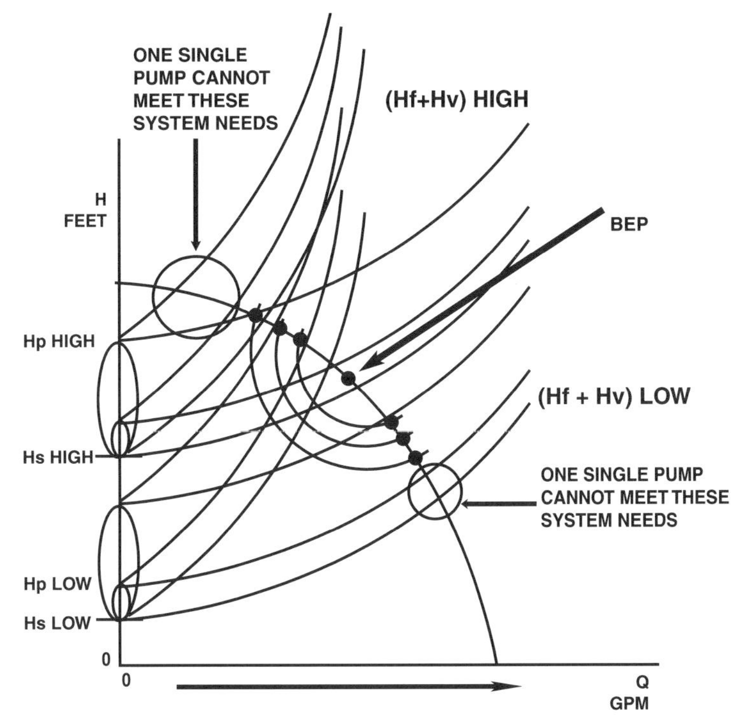

Figure 8–24

liquid from a common system, and discharging the liquid into the same common system. Two pumps running in parallel offer twice the flow at the same head. Pumps in series are two or more pumps where the discharge of one pump feeds the suction of the next pump in series. Two pumps running in series offer twice the head with the same flow. And the combination of the two arrangements offers up to multiples of both factors. First let's consider the arrangement of pumps in parallel.

Pumps in parallel

The system is designed so that two equal pumps are operating together side by side. The system can support the production of both pumps. If the needs of production are reduced, this system can operate with only one pump, simply by removing one pump from service (Figure 8–25).

The curves of pumps 'A' and 'B' individually, and 'A and B' in parallel are seen in Figure 8–26.

Because the system is designed for both pumps running together in parallel, the system curve appears as shown in Figure 8–27.

Here we see something interesting. Because the system is designed for both pumps running in parallel, when only one pump is operating, this pump will run to the right of its BEP. This situation brings it's own peculiar set of implications, not often understood in industry.

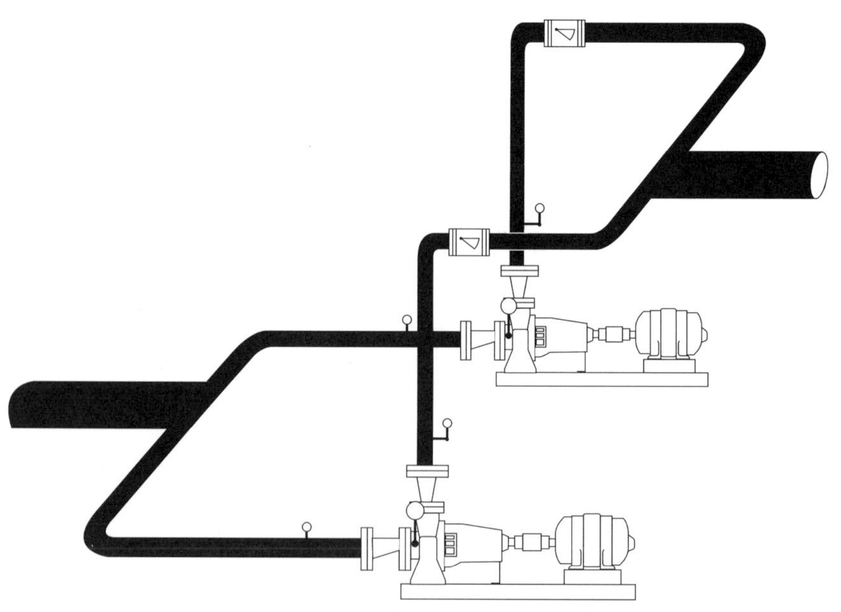

Figure 8–25

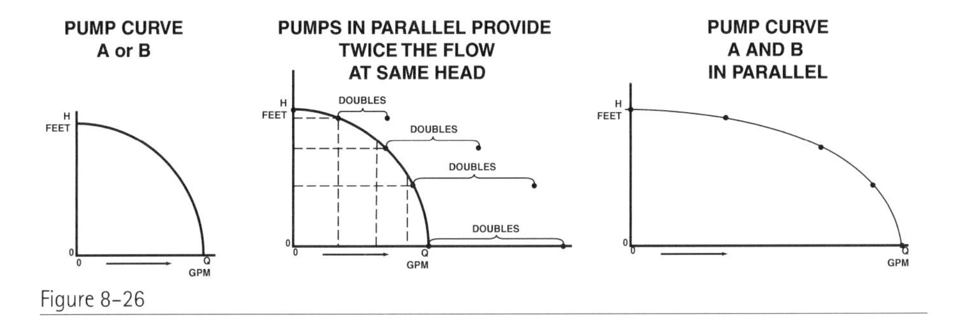

Figure 8-26

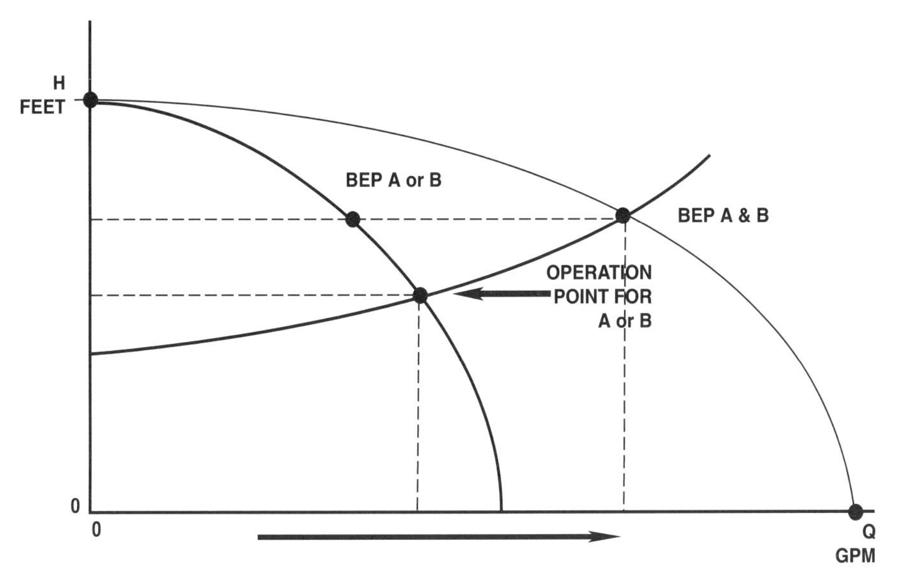

Figure 8-27

Three tips

First, one pump running in a parallel system tends to suffer from cavitation because operation to the right of the BEP indicates that the NPSHr of the pump rises drastically. To survive this condition, you should use dual mechanical seals on these pumps. Dual or double mechanical seals can withstand cavitation better than a single seal. There is a discussion on this in the mechanical seal chapter of this book. Many engineers perceive that parallel pumps are problematic because they appear to suffer a lot of premature seal failure. Parallel pumps deserve double seals even if it's only a cold water system.

The solution is: Parallel pumps should have dual or double seals installed to withstand cavitation when one pump is running solo.

Second, one single pump operating to the right of the BEP indicates that the pump will consume more energy and may require a more powerful motor. For example, if two parallel pumps running together consume 19 horsepower (BHp) of energy, it would seem natural to install a 10-Hp motor on each pump, where the individual consumption would be 9.5 horses each. But operating one pump to the right of its BEP, indicates that this pump might consume 11 or 12 horsepower. Therefore, it might require a 15 horsepower motor installed for running solo. Operating together, the two parallel pumps will only burn 9.5 horses each for a total of 19 BHp.

The solution is: Be prepared to step-up the horsepower on the motor of one solo pump in a parallel system.

Third, you would suppose that parallel pumps are identical, that they were manufactured and assembled together. But it is possible that one pump of the pair is the dominant pump and the other is the runt pump. If you start the dominant pump first in the parallel system, and then decide to add the runt pump of the pair, the weaker pump may not be able to open the check valve. The pump operator perceives that the flow meter on the second pump is stuck or broken. This is because the second pump might be 'dead heading' against a closed check valve, maintained that way by the dominant pump. If this situation exists, it may result in premature failure of bearings and seals, leading maintenance and operations personnel thinking that parallel pumps are problematic.

The solution is: Identify the dominant and weak pump should they exist. To do this, take pressure gauge readings with the pumps running at shut-off head. Verify that the impellers are the same diameter, and that the wear bands and motor speeds are equal. If you can identify one pump in the pair as dominant, always start the weak pump first and then add the dominant pump in parallel with the weaker. The dominant pump coming on stream will push open the check valve. It may be necessary to override a sequential starter.

Once these three points are understood regarding parallel pumps, these pumps give good service in systems that demand more than one single pump can deliver.

Pumps running in series

Let's begin by viewing an arrangement of pumps running in series, followed by the development of the series curve (Figure 8–28).

Series pumps theoretically offer twice the pressure at the same flow (Figure 8–29). The second pump takes the discharge head of the first

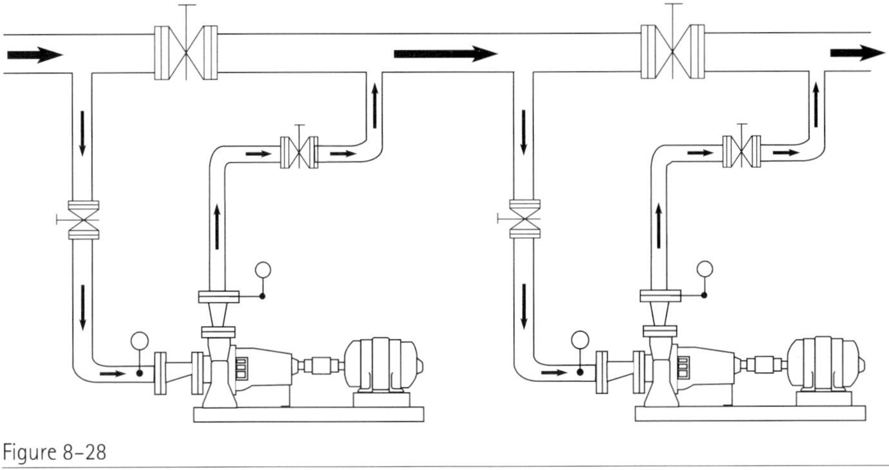

Figure 8-28

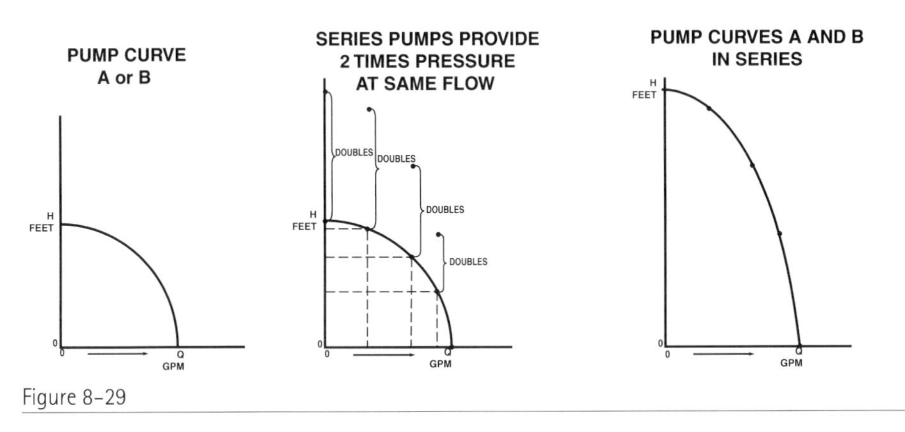

Figure 8-29

pump and jacks up the head again. However, because the system design includes 4 'T' connections, 6 valves, and at least 6 pipe elbows right at the pumps, the actual pressure is not quite doubled because the Hf is significant through the arrangement. The same tips that apply to pumps in parallel, also apply to pumps in series. Depending on the profile of the system curve, one solo pump running in a series arrangement may be running to the left of its BEP or even at shut-off head. If it is running at shut-off head, you don't really have the option of running one solo pump. Use double mechanical seals. It will be necessary to identify and trace the elements of the TDH, and match the TDH to the curve of the pumps running in series.

Combined parallel and series pump operation

Finally, we consider an arrangement of pumps running in combination

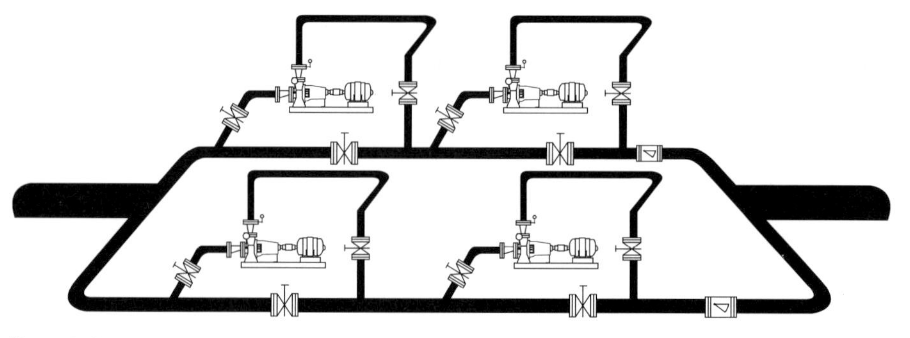

Figure 8–30

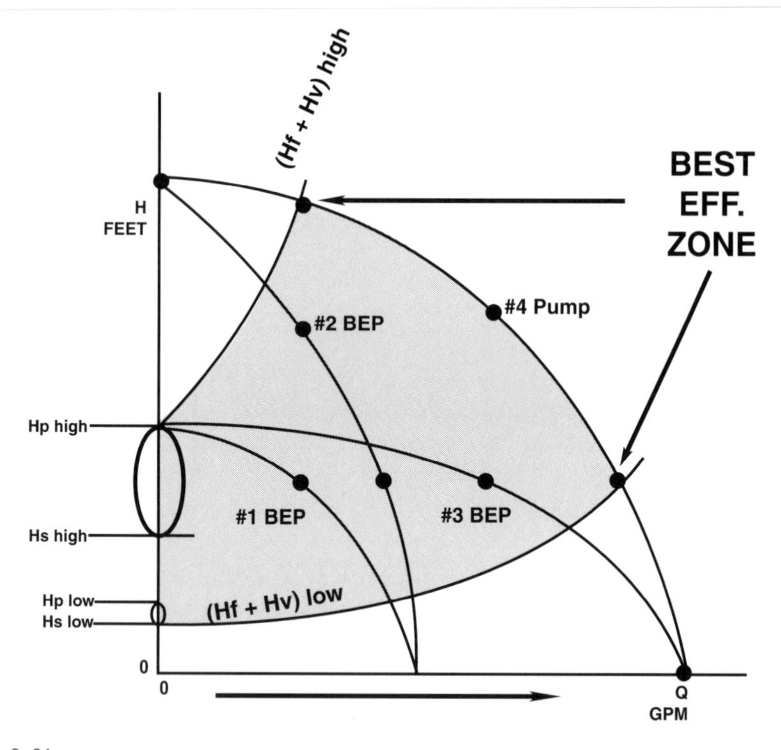

Figure 8–31

parallel and series. Notice that this system design requires 12 gate valves, 2 check valves, 10 'T' connections, and 20 elbows. Because of the high Hf in the area of the pumps, the actual head and flow characteristics may be less than the theoretical characteristics. It appears as in Figure 8–30.

The same previously mentioned critical tips apply, plus one more. Upon observing the system curve, with the pump curves, it appears that the operator can operate any one pump, or any two, or any three or four pumps. Actually there is no option to run three pumps in this arrangement. Any three pumps, by the system design, indicate that you'll be operating two pumps on one side of the system and one pump on the other side. The third pump will not be able to open the check valve with two pumps keeping it closed. So in practice, you can operate any one pump, or any two pumps (with the aforementioned hints from the parallel operation section), or four pumps, but not three pumps.

The curve, shown in Figure 8–31, is indicative of this operation.

Shaft Deflection

Introduction

Along with the sounds, evidence and signs of cavitation, there is a broad range of other information and signals available to the maintenance mechanic. Almost all mechanics have seen the gouge and scratch marks, and signs of heat on the pump when disassembled in the shop. Sadly, most mechanics are never trained to interpret these marks.

This brings us to failure analysis of the pump, or performing an autopsy on a broken pump. You must stop throwing away used and worn pump parts, or sending them to the machine shop. This action destroys the evidence needed to repair and resolve pump problems. There are too many mechanics wasting their careers changing parts and not really repairing anything.

Let's begin with a discussion and explanation on how a volute centrifugal pump works.

60° and 240°

The volute type pump has its impeller mounted eccentrically within the volute. The degree of eccentricity governs the pressure that the pump can generate. If the impeller were concentric inside the volute, or equidistant, the pump would generate flow, but not much pressure or head (Figure 9–1).

The impeller throws the liquid against the volute wall at a constant speed, the speed of the electric motor. The internal diameter of the volute wall converts the velocity into head or pressure (Figure 9–2).

See Table opposite for what is happening inside the pump around the internal volute wall.

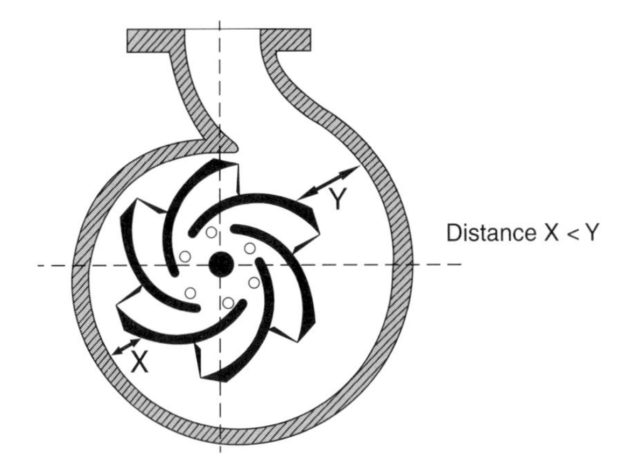

Figure 9–1

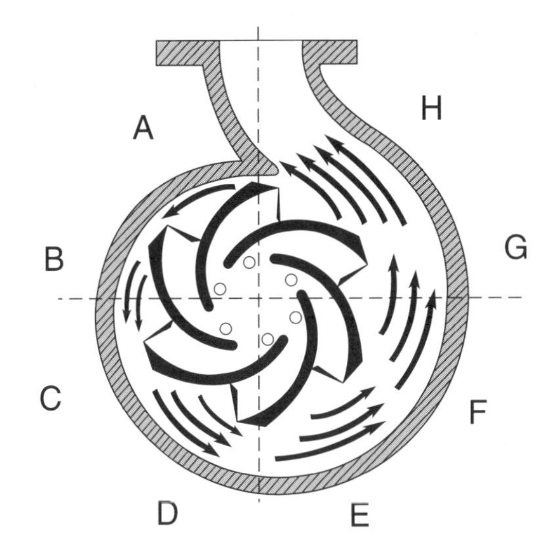

Figure 9–2

AT POINT	PRESSURE	VELOCITY	AREA
A	LOW	HIGH	LITTLE
B	HIGHER	LOWER	MORE
C	HIGHER	LOWER	MORE
D	HIGHER	LOWER	MORE
E	HIGHER	LOWER	MORE
F	HIGHER	LOWER	MORE
G	HIGHER	LOWER	MORE
H	THE MOST PRESSURE	THE LEAST VELOCITY	THE MOST AREA

Presssures are Equal.

Figure 9–3

With the pump running at its Best Efficiency Point, and all valves in the system open, the factors of pressure, velocity, and area are in harmony at all points around the volute. All radial loads are in equilibrium (Figure 9–3)

If a discharge valve should be throttled (increasing the resistance head, Hf), the pressure gradients around the volute would tend to equalize toward discharge pressure. In a worst-case scenario, if a valve should close completely, the pressures around the volute would become discharge pressure. The pump would move to the left of its BEP on the curve. The velocity would become zero because no fluid is moving through the pump. The only remaining variable is the area, which is greater through the E-F-G-H arc of the volute circle (Figure 9–4).

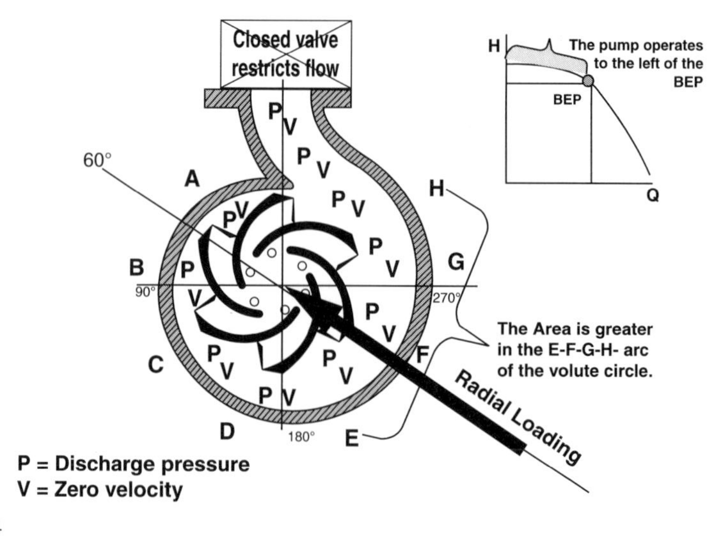

Figure 9–4

With pressures equal and more area in the E-F-G-H arc of the volute circle, a tremendous radial force is created that will distort and deflect the shaft toward a point approximately 60° around the volute from the cutwater. This radial force can destroy the mechanical seal or packing rings, bearings, and deform and even break the shaft. The evidence would be rub or scratch marks around the circumferences of close tolerance rotary elements, such as the outer diameter on open or semi open impellers (see Point A in the next illustration, Figure 9–5), the wear rings on closed impellers (see Point B, next illustration), the shaft or sleeve at the restriction bushing in the bottom of the seal chamber or stuffing box (see Point C), or on the posterior of the mechanical seal (Point D).

The scratch marks on the circumference of these close tolerance rotary parts will correspond to scratch marks on close tolerance stationary parts at approximately 60° around the volute from the cutwater. These marks will be visible on the back plate with open impellers, or on the wear rings of pumps with enclosed impellers, or the ID bore of the restriction bushing at the bottom of the seal chamber where the shaft passes through, or the ID of the seal chamber bore at the back end of the mechanical seal (Figure 9–6 and Figure 9–7, next page).

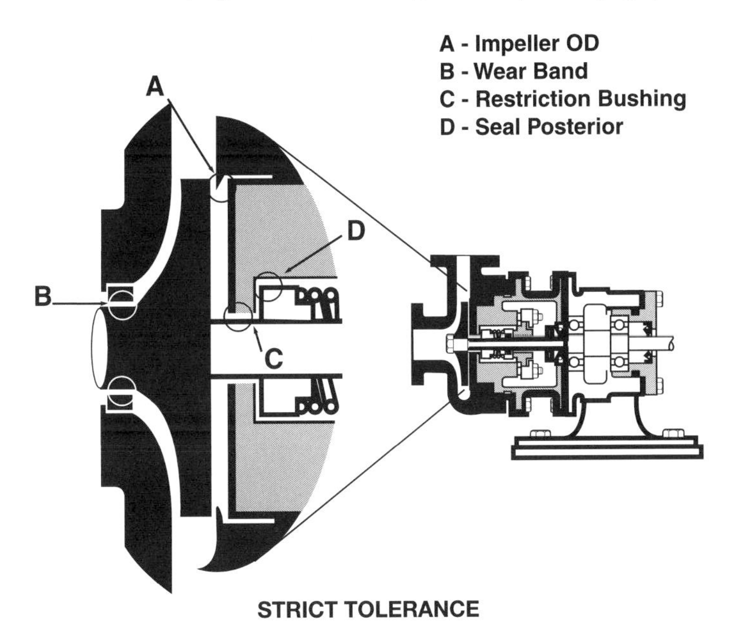

Figure 9–5

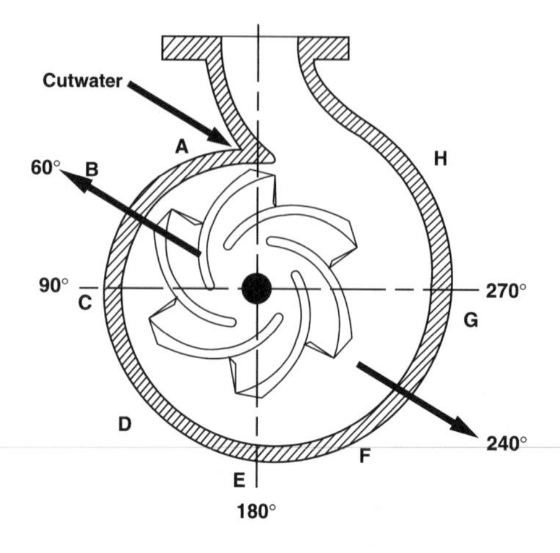

Figure 9–6

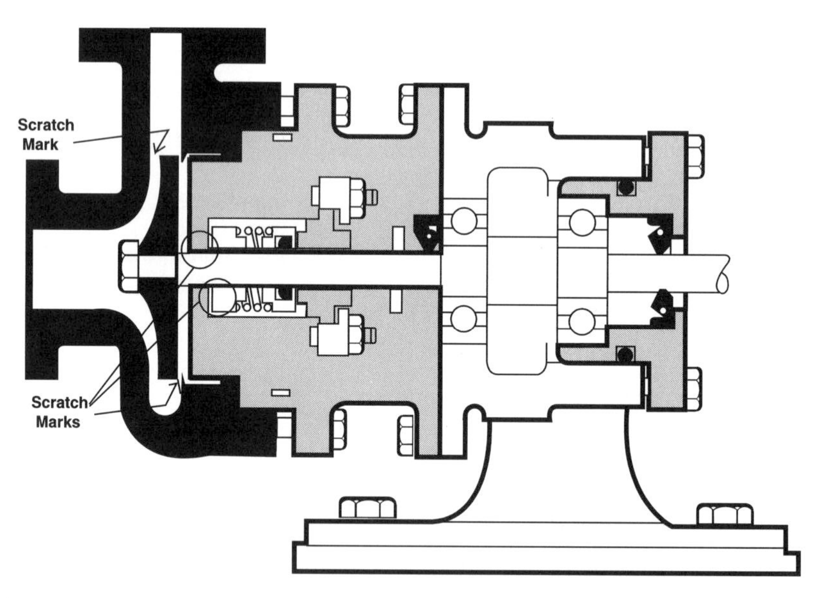

Figure 9–7

The other case is when there is too much flow through the pump. The pump is operating to the right of the BEP on its curve (Figure 9–8). The same problem occurs, but now in the other direction. With the severe increase in velocity through the pump, the pressures fall dramatically in the E-F-G-H arc of the volute circle (Bernoulli's Law says that as velocity goes up, pressure comes down). Now the shaft deflects, or even breaks in the opposite direction … at approximately 240° around the volute from the cutwater.

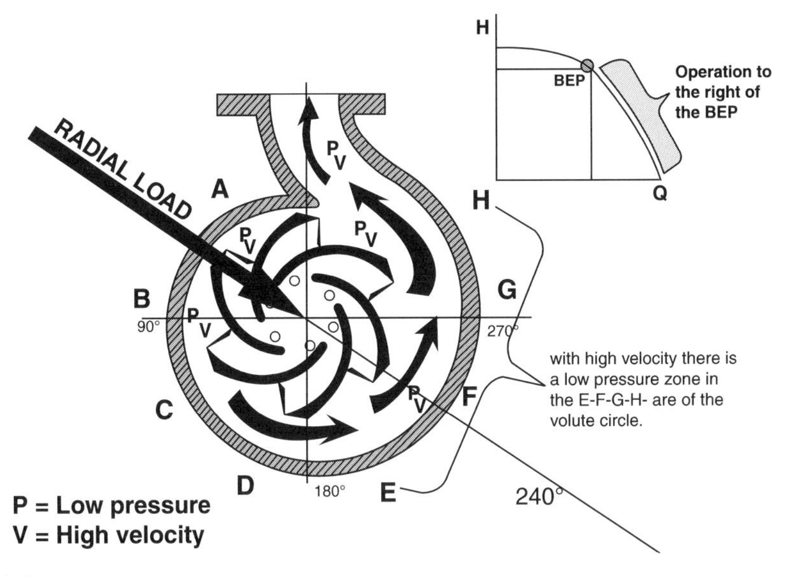

Figure 9-8

Depending on the pumps available in your manufacturing or process plant, you will experience:

- Broken Shafts.
- Premature Bearing Failure.
- Premature Mechanical Seal Failure.
- Premature Packing Failure.
- Worn and Damaged Shaft Sleeves.
- High Maintenance Costs on your Pumps.

You can't begin to resolve problems in your centrifugal pumps, bearings and mechanical seals, until you learn the numbers 60°, and 240°, with respect to your pump cutwater.

Operation, design and maintenance

Once again, when the pump is operating at its BEP, all forces within the volute (velocity, pressure, and the area exposed to velocity and pressure) are in equilibrium and harmony. The only load on the bearings is the weight of the shaft. The pump, the mechanical seal, and the bearings will run for years without problems. When problems arise that cause high maintenance costs with the pump (remember that seals and bearings are the principal reason that pumps go into the shop) these problems normally originate from one of three sources:

- Problems induced by Operations.
- Problems induced by Design.
- Problems induced by Maintenance.

Let's analyze the evidence that pump mechanics have seen so many times. Consider the difference between a deflected shaft and a bent shaft.

A bent shaft is physically bent and distorted. Placing the shaft into a lathe or dynamic balancer and rotating it will reveal the distortion. If a bent shaft is installed into a pump and run, it will fail prematurely, leaving evidence and specific signs on the circumference of close tolerance stationary parts around the pump's volute circle. The shaft will exhibit a wear spot on its surface where the close tolerance parts were rubbing.

A deflected shaft is absolutely straight when rotated in a lathe or dynamic balancer. The deflection is the result of a problem induced either by operation or system design. The deflected shaft also will fail prematurely in the pump, leaving similar, but different evidence on the close tolerance rubbing parts in the pump. The next two pictures show how a bent shaft appears when rotated 180 degrees (Figure 9–9 and Figure 9–10).

The basic difference between a bent shaft and a deflected shaft is the following. A bent shaft spinning inside close tolerances leaves a scratch mark around the circumference of stationary elements corresponding to a damaged spot on the shaft. A deflected shaft spinning within close

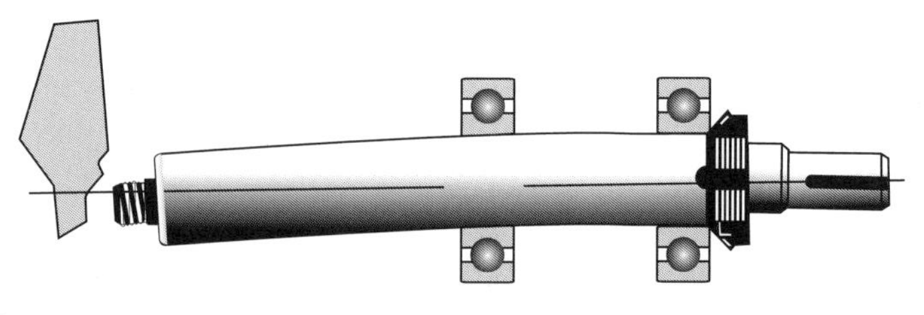

Figure 9–9

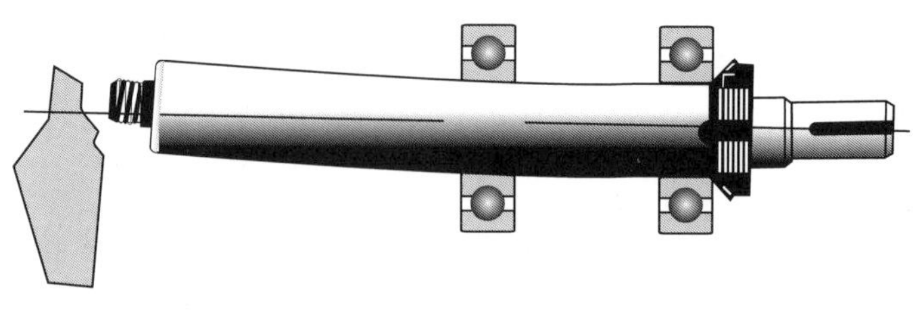

Figure 9–10

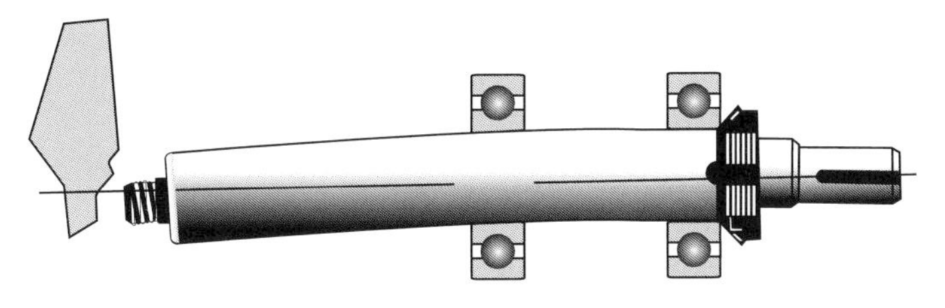

Figure 9–11

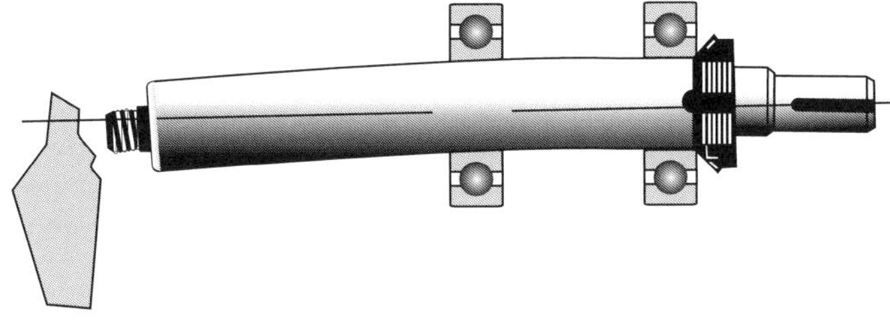

Figure 9–12

tolerances leaves a scratch or gouged circle around the rotary element, and a gouged or damaged spot on the stationary elements. It is absolutely necessary to distinguish and recognize these significant differences.

If the pump is put into service with a bent or unbalanced shaft assembly, its premature failure can be traced to inadequate maintenance practices. The evidence does not lie. However, if the premature failure leaves evidence of a deflected shaft, this would be an operations or design failure. All too often, the mechanic is blamed. The two pictures above show how a deflected shaft appears when rotated 180 degrees (Figure 9–11 and Figure 9–12).

Shaft deflection is the result of an external radial load. The external radial loading originates with the pump operator or process when the pump runs away from its best efficiency point on the curve. The resistance to deflection is a function of the shaft's overhang length and its diameter. The deflection resistance, also called the flexibility factor, is known as the L/D factor.

The L/D indicates length/diameter. Because pumps are manufactured with certain dimensional standards (ANSI, API, DIN, and ISO), the L/D factor can and should be specified at the moment of specifying the pump. The design engineer could request that the pump manufacturer quote a pump based on its flow, head, metallurgy, and L/D factor, awarding bonus points for a low L/D, indicating a high deflection resistance. The high deflection resistance is an index of how far the pump can be run away from its BEP on the curve without damaging the mechanical seal and bearings.

Rarely do design engineers request the L/D factor in their quotes. Some engineers don't know they have the option. Most pumps are bought based on price, and because a high deflection resistance (low L/D Factor) indicates a larger diameter shaft with oversized bearings; these type pumps don't normally win a competitive bid process.

If you suspect, or know, that you have a deflected shaft, or know that standard operating procedure in your plant requires controlling the flow in the pipes by opening and closing valves, then you have three options to reduce shaft deflection:

- Use a larger diameter shaft.
- Use a shorter shaft (this may affect the motor mounts, and/or piping mounts).
- Change the shaft metallurgy (this will change the elasticity modulus and may even start a round of galvanic corrosion).

Increasing the shaft diameter is the most logical solution. This can be done with some pump models by simply replacing sleeved shafts with solid shafts, or by increasing the diameter of the solid shaft with a small modification to the seal chamber bore. With the pump disassembled on the shop table, the mechanic can identify the source of the problem in the pump.

Signs of shaft deflection

Most pumps have tight tolerances in the following rotary elements:

- The OD of the blades on open and semi open impellers.
- The wear bands on pumps with enclosed impellers.
- The shaft under the restriction bushing at the bottom of the stuffing box or seal chamber.
- The OD of the posterior end of the internally mounted mechanical seal.

These tight tolerance rotary elements have corresponding tight tolerance stationary elements. These are:

- The internal volute wall and/or back plate on pumps with open and semi open impellers.
- Stationary wear band bores on enclosed impellers.
- The restriction bore at the bottom of the stuffing box or seal chamber where the shaft passes through.

- The seal chamber internal bore corresponding to the posterior of the mechanical seal OD.

Interpreting the evidence

Let's interpret the physical evidence that you might see at these close tolerances and their source. To begin:

1. You might see gouge or wear marks all around the circumferences of close tolerances on the rotary elements, and a corresponding wear spot at approximately 60° from the cutwater on the stationary elements.
    - This would be induced or caused by operation, if the plant operators strangle valves to control production flow. Any other discharge flow restriction (clogged filter, pipe obstruction, or un-calibrated automatic valve) would produce the same evidence. Talk with the plant engineer about this situation and show him the evidence on the pump.
    - This could also be induced by design, if the pumps are oversized, or by high velocity and friction head in discharge piping of inferior diameter.
    - This could be induced or caused by maintenance in cases where the mechanic installs a check valve in reverse, or uses inadequate practices when rebuilding valves, cutting and placing flange ring gaskets at pipe joints, or exchanging and replacing incorrect valves. Our recommendation is to always use good maintenance practices.

AUTHOR'S NOTE

Of the three sources of problems, design, operation and maintenance, the mechanic is really responsible for a small part. The truth is that the majority of pump problems begin with changes to design, and plant operations after the system was commissioned.

- If the condition should be occasional, the solution could be to install a variable speed motor. If the condition is permanent, the solution could be to reduce the impeller diameter, replace the pump, or increase the diameter of the pipe. If normal operations require living with the condition, then increase the diameter of the pump shaft to improve the L/D factor.

2. You may see the same evidence all around the circumference of the close tolerance rotary elements, with gouge or wear spots on the stationary elements at about 240° from the pump cutwater.

   - These marks are caused or induced by operations or by design. This evidence is revealed when operating the pump too far to the right of the BEP on its curve. Perhaps the pump is inadequate and doesn't meet the flow and head requirements of the system. It could also be that there is a loss of resistance in the discharge piping. A big hole in the discharge piping could present the same evidence.
   - If you must live with this condition, you need to increase the diameter of the shaft to improve the L/D factor and deflection resistance.

3. If you see the same evidence, gouge and wear marks around the circumference of close tolerance rotary elements, and spots or arcs on the close tolerance stationary elements at about 180° from the cutwater, or straight down:

   - This would be a problem induced by inadequate design, caused by pipe strain probably in a high temperature (thermal expansion) application. The volute of the pump and the stationary elements are growing up from the floor due to thermal expansion, against the rotary elements. You need to speak with the plant engineer and show him the evidence. A possible solution is to change your ANSI standard pump for a 'High Temperature' or API design in this application.

See the following graphs, Figure 9–13, depicting thermal expansion. The picture on the left shows an ANSI pump where thermal growth is straight up from the base. On the right we see a high temperature pump where thermal growth occurs 360 degrees around the volute.

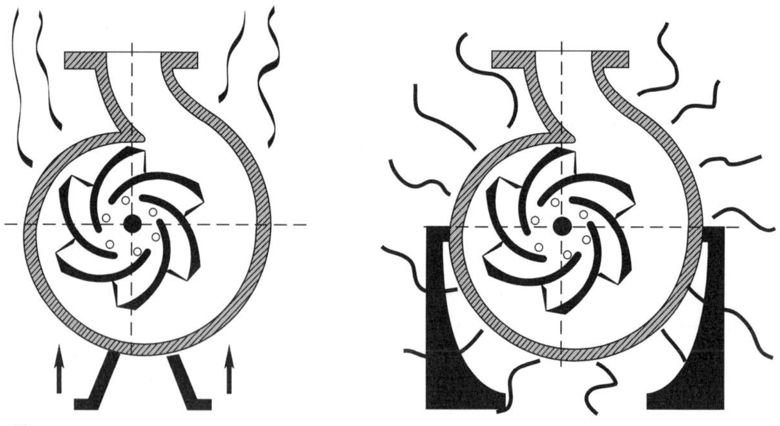

Figure 9–13

The coefficient of thermal expansion of 316 stainless steel is 9.7×10^{-6} in/in per degree Fahrenheit. The metric equivalent is 17.5×10^{-6} mm/mm. per degree Centigrade. See the next Table and note the expansion on a pump whose centerline is 10 inch above its base.

ΔTEMPERATURE F°	ΔTEMPERATURE C°	THERMAL EXPANSION Inches	THERMAL EXPANSION Millimeters
100 °F	55 °C	0.0097	0.245
200 °F	110 °C	0.0190	0.490
300 °F	165 °C	0.0291	0.735
400 °F	220 °C	0.0388	0.900
500 °F	275 °C	0.0485	1.230
600 °F	330 °C	0.0582	1.470

As you can see, the pump casing will grow against its shaft almost 0.030 inches with an increase of 300°F. There are many tolerances in a pump that are tighter than 0.030 inches. This means a rotary element will scrape and rub a stationary element.

You may even see the same evidence of gouges and wear around the circumference of strict tolerance rotary elements, leaving a corresponding spot on the stationary elements at any other point around the volute circle of the pump.

This condition is probably misalignment, indicating a maintenance problem. The mechanic should be trained to correct this. Follow correct alignment procedures, as well as correct bolt torque procedures.

Inspect gasket surfaces for knots and irregularities. Look for bent dowel pins and misaligned jack bolts, dirt and any other factor that might lead to misalignment.

Next we'll discuss evidence marks and prints that are different, but to the untrained eye, they may appear the same. You may see a spot or arc of wear and gouging on the rotary elements, and a circumferential wear circle on the bore of the close tolerance stationary elements. This is a maintenance-induced problem. This is the sign of a physically bent shaft, or a shaft that is not round, or a dynamic imbalance in the shaft-sleeve-impeller assembly. The solution is to put the shaft on a lathe or dynamic balancer, verify its condition, and correct before the next installation.

The next condition and physical evidence we'll mention is rare, but we need to cover it in case you should ever see it. You might see scratch and gouge marks all around the circumference of strict tolerance rotary element ODs, and stationary element bores alike. This condition and marks is evidence of a **'Lack of Control'**. It could be from any of the aforementioned reasons up to this moment, and even including vibration, damaged and misapplied bearings.

The problem could be maintenance, operation, or design, or a combination of any or all these factors. In all honesty, you should never see this set of evidence marks because it indicates a lack of control. Now because the mechanic cannot control operational problems or design problems, the first phase to correct this situation is to control the mechanical maintenance factors, like alignment, proper bolting and torque sequences, be sure shafts are straight and round, and dynamically balance all rotary components. Reinstall the pump and wait for the next failure. Once the maintenance factors are under control, there should appear a clear vision and path to resolve any operational and/or design weaknesses.

The sweet zone

Consider the following graph, Figure 9–14. Radial loading on the shaft rises if the pump is operated too far to the left or right of the best efficiency zone. Another interpretation of the same concept is to say that the maintenance and problems rise when the pump is operated away from its BEP. Many pumps have a rather narrow operational window. These pumps can be very efficient if they are correctly specified and operated. This is discussed completely in Chapters 7 and 8 Pump Curves and System Curves.

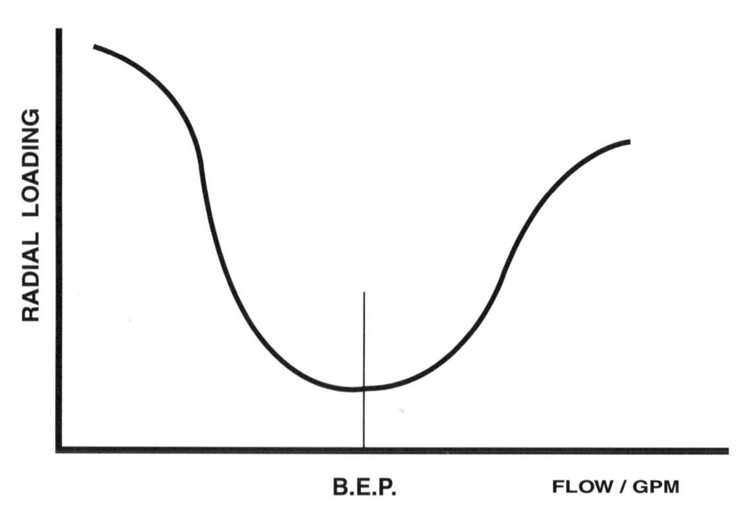

Figure 9–14

The dual volute pump

Some pump manufacturers use a different tactic to expand the operating window of their products. This is why the dual volute pump exists. The dual volute pump is designed to operate over a wide range of flows and heads.

In the casting procedure of the dual volute pump, a second cutwater is designed 180° opposite the first cutwater and another volute channel added. This way, all areas exposed to velocity and pressure around the volute casing are equal. All forces around the volute are equalized or cancelled (Figure 9–15).

The second cutwater and volute channel create an additional obstruction to the flow through this pump, and for that reason it robs some energy from the fluid. This will reduce the efficiency to a small degree, but the operational range of these pumps is quite broad as you can see in the graph. The dual volute design casing is an optional cost accessory with some pump companies, and offered standard with others.

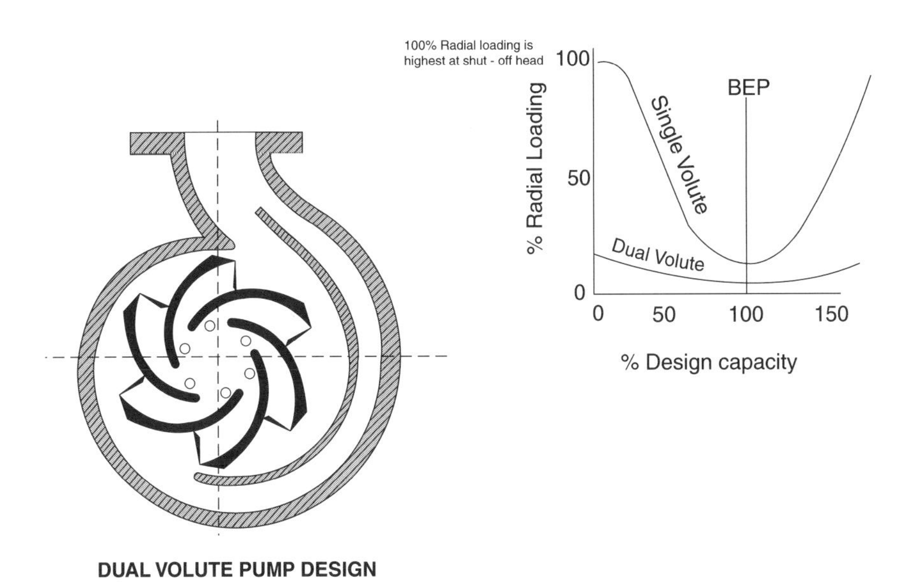

Figure 9–15

Pump and Motor Alignment

Introduction

Pump shaft and driver shaft alignment is very important for long useful equipment life, and to extend the running time between repairs. Besides, good alignment reduces the progressive degradation of the pump.

AUTHOR'S NOTE

Why do we use the word driver? We tend to think that pumps are powered by electric motors. However, some pumps are powered by internal combustion engines, or with turbines or hydraulic motors. Not always are pumps and drivers connected with a direct coupling. Some pumps are coupled through pulleys, chain drives, gearboxes or even transmissions.

If the pump shaft and impeller assembly were perfectly balanced and aligned, it would rotate in a perfect orbit around the shaft centerline. This condition is practically impossible. There is always some imbalance in the shaft and impeller assembly due to its casting and machining process, and perfect alignment doesn't exist. Because of this, the shaft spins eccentrically around the centerline. We could call this movement 'eccentric rotation'. The implications of a pump exhibiting rotary assembly imbalance (eccentric rotation) include:

- Excessive running noise.
- Vibration and excessive loads on the bearings causing premature failure.
- Rapid wear of the coupling and eventual premature failure.
- Premature packing or mechanical seal failure.

- Wear and rubbing between close tolerance rotary and stationary elements in the pump leading to their failure.
- Premature driver bearing failure.
- Increased energy consumption.
- Excessive operating temperatures and lubricant failure.

One of the most important and least considered points of correct alignment is the relationship with the power transmitted from the motor to the pump. An almost perfect alignment (0.003 inch) with an adequate and new coupling transmits almost 100% of the motor's power (there will always be some small losses). The Pump performance curve identifies the BHP or brake horsepower required for the pump to perform at its duty point.

The next graph (Figure 10–1) indicates the expected continuous running time of rotating equipment with increasing misalignment. As you soon see, the alignment improves and so does the service time.

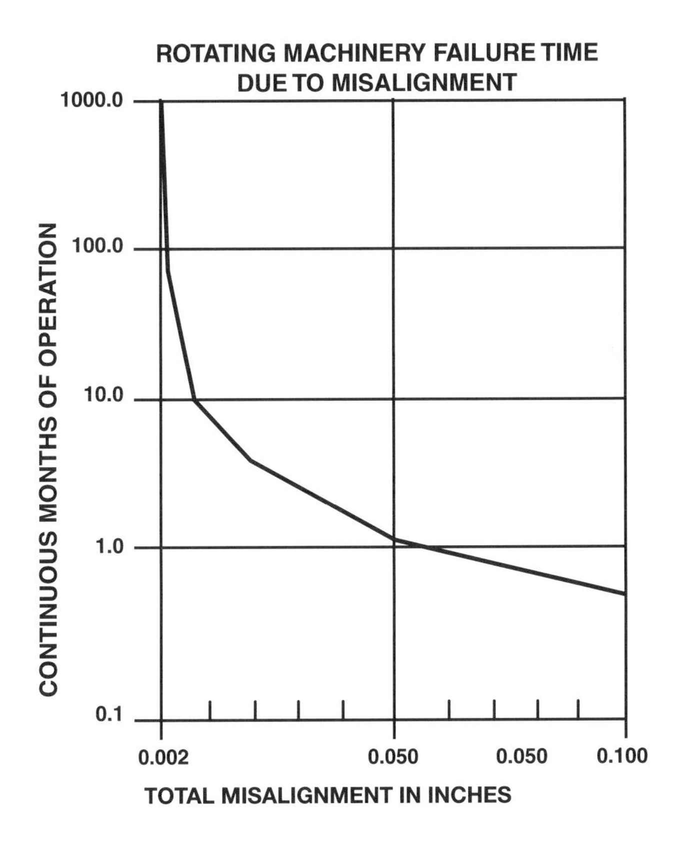

Figure 10–1

Types of misalignment

There are two basic types of misalignment, angular and parallel. Within each of these basic types of misalignment there are combinations of both. These are the most common combinations:

- Vertical/angular misalignment (Figure 10–2
- Vertical/parallel misalignment (Figure 10–3)
- Horizontal/angular misalignment (Figure 10–4)
- Horizontal /parallel misalignment (Figure 10–5)
- Combined angular and parallel misalignment

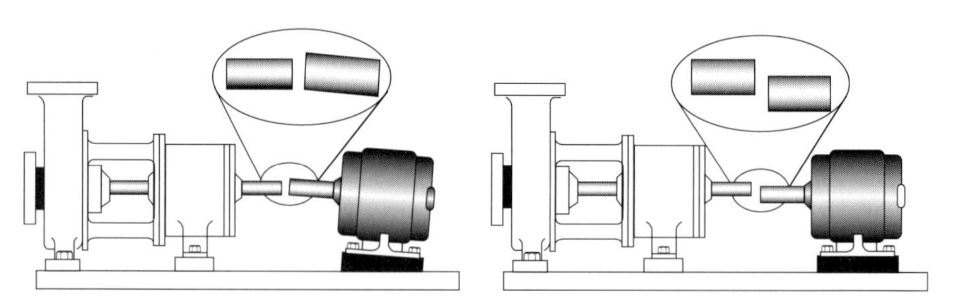

Figure 10-2 and Figure 10-3 Side view of vertical/angular and vertical/parallel misalignment

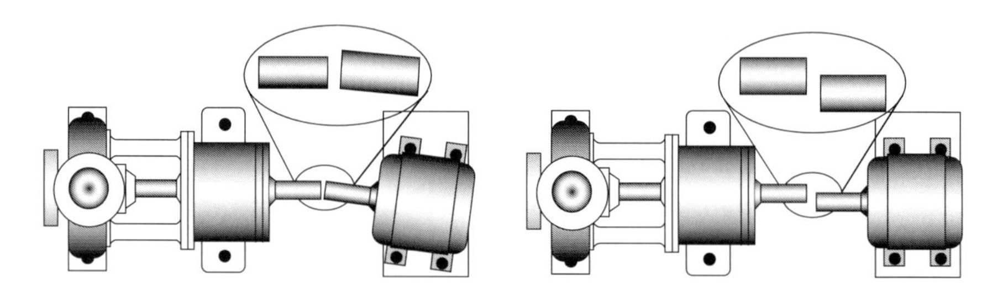

Figure 10-4 and Figure 10-5 Top view of horizontal/angular and horizontal/parallel misalignment

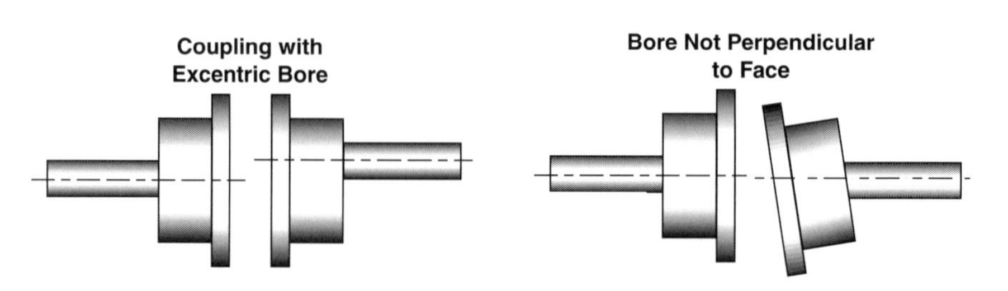

Figure 10-6 and Figure 10-7 Misalignment can be transmitted through the couplings and coupling faces.

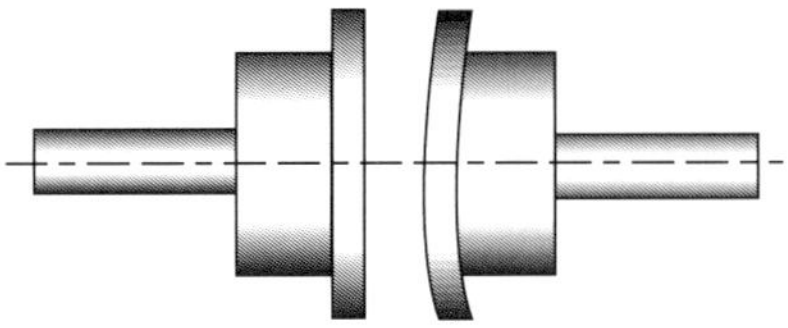

Distorted Coupling Face

Figure 10-8 Misalignment can be transmitted through the couplings and coupling faces.

Alignment techniques

There are a variety of shaft alignment procedures. The configuration and size of the equipment determines the best alignment method. Generally the driver or motor should be aligned to the pump. The motor shaft centerline should be shorter and brought up to the pump shaft centerline with shims or spacers. The pump is generally fixed and attached to the suction and discharge piping, so it is almost impossible to move. The volute casing aids in supporting the piping, so it should be fixed to a solid foundation without shims, jack bolts, or supports. Verifying the alignment of running equipment is critical to maintain the correct operation and reduce downtime.

Most established alignment procedures call for the use of precision dial indicators to correct misalignment. Gaining popularity in industry is laser alignment technology. We'll cover this shortly. Among the most popular methods of alignment are:

- Reverse Dial Indicator alignment.
- Rim and Face alignment.
- Straight Edge alignment.
- Laser Alignment.

Reverse dial indicator alignment

This is the most popular method used in industry today because the investment in equipment is moderate and its effectiveness is proven. This method uses two dial indicators, one on the pump shaft and the other on the motor shaft.

Sometimes in practice the dial indicators are mounted on the couplings, but it is best to mount and fix the indicators onto the shafts because the couplings may be eccentric to the shaft centerlines. Rotate the shafts and obtain the displacement readings. Project these readings graphically or mathematically to the motor base to determine the adjustments required, and the spacing shims under each foot.

Rim and face alignment

This method is most useful when only one of the shafts can be rotated for the alignment procedure, or when the two shaft ends are very close to each other. Obtain the displacement readings with the dial on the rim (OD) of the coupling and the coupling face. Project these readings mathematically or graphically to the motor base to determine the required adjustments and shims for each foot. This method is not as precise and may have a built-in error, if the coupling center is eccentric from the shaft centerline.

Laser alignment

Laser alignment systems use a transmitter and receiver. The system has a laser diode and a position sensor on a bracket mounted on one shaft that emits a weak and safe radio-tagged beam of light. The light ray is directed toward the other bracket on the other shaft with a reflecting prism that returns the ray back toward the first bracket into the position sensor eye.

One shaft is rotated to determine the vertical and horizontal readings as in the other alignment techniques.

The shaft alignments are automatically entered into a small computer that calculates the relative required movements needed at the motor base to align the two shafts. See Figure 10–9.

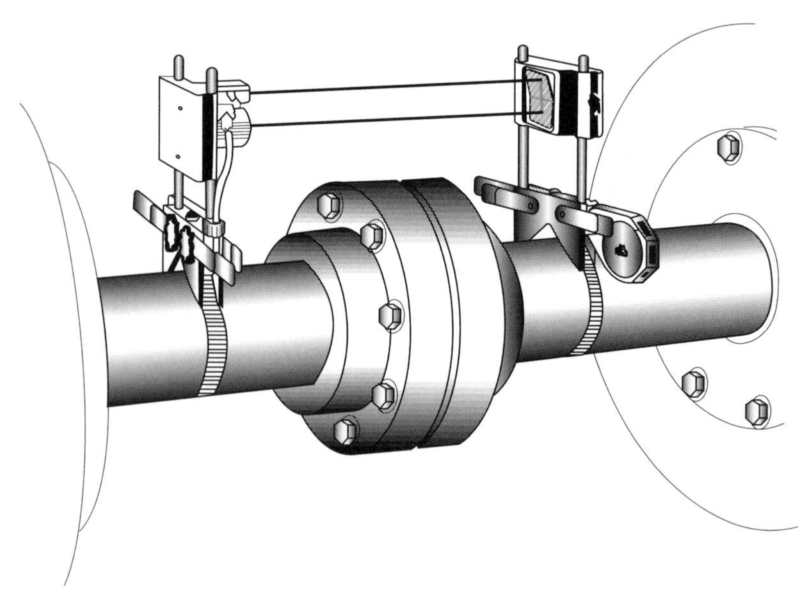

Figure 10–9

General observations on the alignment process between shafts

1. The alignment procedure should be repeated at various intervals to identify installation errors and compensate for equipment operation. This is the way to assure long equipment life. It is recommended to go through the alignment procedure and make corrections in the following stages:

   - At Pump Installation: Be sure the motor shaft centerline is below the pump shaft centerline so that it can be shimmed upward. Make sure the motor mount boltholes have sufficient play to allow for some lateral adjustment. Many pumps and motor assemblies are shipped from the factory on a common channel iron base plate. The manufacturer alleges that they are already aligned at the factory. You need to verify and correct this alignment in all cases.
   - After connecting the piping and accessories: Before starting the pump, repeat the procedure after all associated connections have been made. If there is a marked difference, the problem may be pipe strain distorting the pump casing through the suction and discharge nozzles. This situation should be resolved with the installation contractor or pipe fitters. Not correcting this situation is sure to bring future maintenance problems from misalignment.
   - Hot alignment: Allow the equipment to run for three or four hours and come up to operating temperature, then shut-off the pump and repeat the alignment procedure with the equipment hot.
   - Running alignment: After the pump has been running for a week or ten days, perform an alignment check to verify that the equipment is not suffering pipe strain or binding from thermal growth.

2. The base and cement foundation should be examined to verify a correct installation. The pump and motor assembly should rest on a common base.

   - The base should be sufficiently strong to withstand the machinery weight and minimize vibrations. Five times the mass is the rule. If the pump, motor, and base plate weighs 1,000 lbs, the foundation should weigh at least 5,000 lbs.
   - The base should be level and flat.
   - The base should be the proper size. This varies according to its size and weight. It should have enough free adjacent space to perform maintenance, alignment and proper cooling.

3. The grout should be the correct type for the climate and application temperature, speeds, and chemical nature.
   - Its function is to absorb the vibrations generated by the motor and pump.
   - It should contain aggregate or epoxy.
   - It should be applied strictly according to manufacturer's recommendations.
   - It should be inspected for fractures, crumbling, and separation every six months to identify conditions affecting the equipment alignment.
4. Bases
   - The driver or motor shaft should be level and parallel with the base.
   - Shims should be free of dirt and corrosion. They should be replaced from time to time because they can become deformed with time and weight.
   - Bases should be inspected for corrosion and corrected if necessary.
5. The Motor
   - During the alignment procedure, follow your plant lockout/tagout procedure to prevent accidents.
   - Motor sleeve bearings require limiting the axial play.
   - Study the coupling manufacturer's instructions to assure the proper spacing between the faces. The spacing is relative to the motor size.
6. Dial indicators
   - During the alignment it is important to note the direction of the indicator movement. Beginning at 0.000 inches, a movement in a clockwise direction is a positive reading. A counterclockwise movement indicates a negative reading; see Figure 10–10.
   - Rotating the shaft and dial 360°, the left lateral reading plus the right lateral reading should equal the sum of the superior and inferior readings.
   - The indicator readings at the end of the rotation should be the same as the readings at the beginning of the rotation.

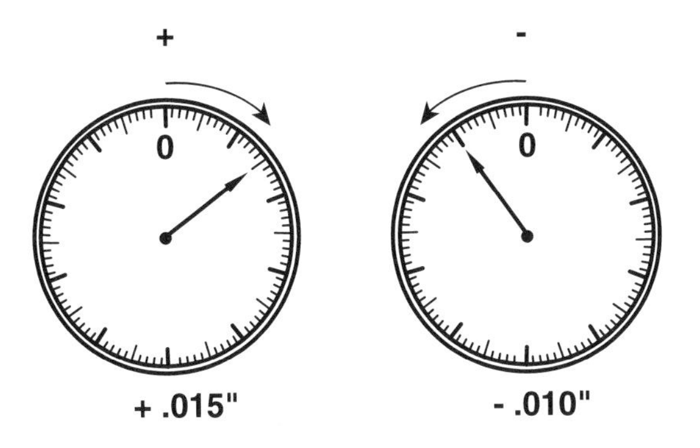

Figure 10–10

7. Shims
   - Spacer shims should be made of 304 stainless steel, except with chlorine and hydrochloric acid service. In these services, use Mylar shims to resist corrosion.
   - It is best to use the thickest shim possible instead of numerous thin shims, which can suffer from compression. Never stack more than 3 shims under an equipment foot.
   - Measure shims to verify their thickness and tolerance, especially thin shims (those less than 0.005 inch).
   - Avoid the use of shims with the thickness stamped on the shim face.
   - Use shims large enough to completely cover the equipment footprint.
   - Avoid rust, scratches, gouges, creases, indentations, hammer blows and dirt.
   - Install the shims sliding them under the machinery footprint, until contact is made with the anchor bolt. Then move the shim back away from the bolt shaft to avoid interference with the threads and to assure tolerance.
8. Alleviate any possible pipe strain, a force imposed by the piping that can distort the pump casing.
   - Pipe strain is normally caused by misalignment between the piping and the pump nozzles, improper pipe supports, or thermal expansion in the system.
   - Don't connect the piping to the pump until the cement base and grouting is fully cured, and all foundation bolts are tightened.

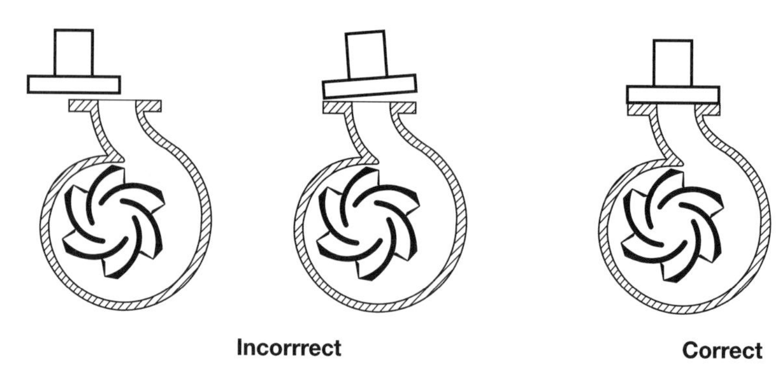

Figure 10–11

- Bring the pipe to the pump and adjust it to the pump. Don't adjust the pump to the piping (Figure 10–11).
- To verify pipe strain, place dial indicators on the shaft and watch for horizontal and vertical movement. Unite the flanges one at a time continually observing the indicator readings. In general the indicator readings should not exceed 0.002 inches (Figure 10–12).

9. Correct Soft Foot. Soft foot exists when one of the four machinery feet is not level with the base. When the base bolts are tightened with soft foot, the effect can distort and misalign the pump casing.
   - To check for soft foot, place a dial indicator onto the machinery foot, and loosening the base bolt. If the indicator moves more than 0.002 inches, the foot is soft and it should be corrected. Go through the same procedure on the remaining feet one at a time.

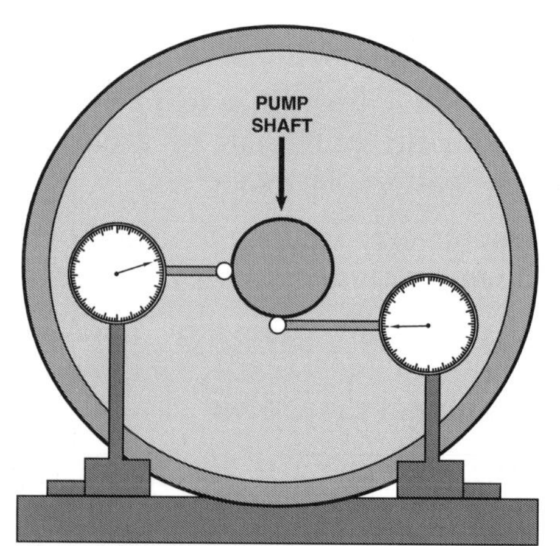

Figure 10–12

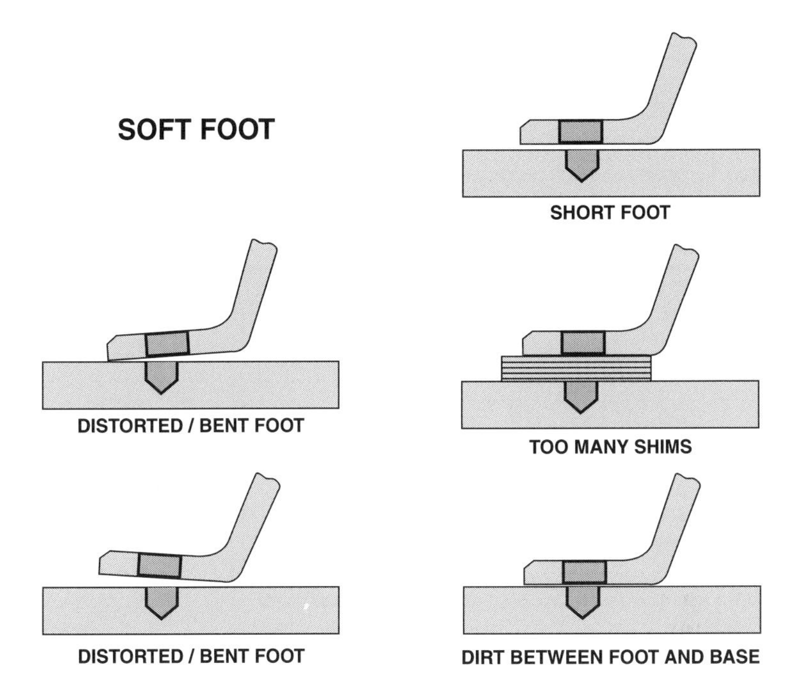

Figure 10–13

- To correct soft foot, place shims under the foot in the thickness corresponding to the movement of the dial indicator.
- If the foot inclines from either the outer or inner border, it will always rise upon loosening the base bolt, and correct alignment will be almost impossible. It will be necessary to re-machine all four feet to achieve parallelism between them.

10. Check for indicator bar shaft deflection.
    - This deflection is due to the weight of the indicator dial.
    - Mount the dial indicators on the equipment in the same manner and distance required to perform the alignment procedure.
    - Start straight up at the top of the shaft and rotate 180° down to the bottom.
    - Note the indicator readings.
    - This deflection can be corrected easily during the alignment. For example, with the indicators in the upper position on the shaft, instead of starting at 0.000 inches, mark the positive value of the deflection of the bar determined in the previous step, and then rotate the shaft 180° to the bottom. Now the indicators will read 0.000 inches.

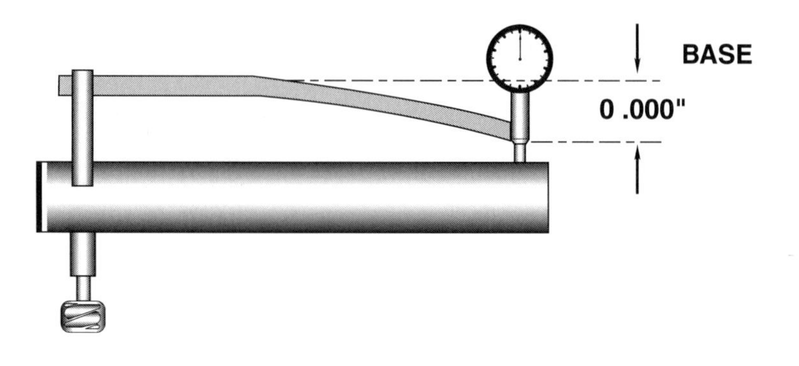

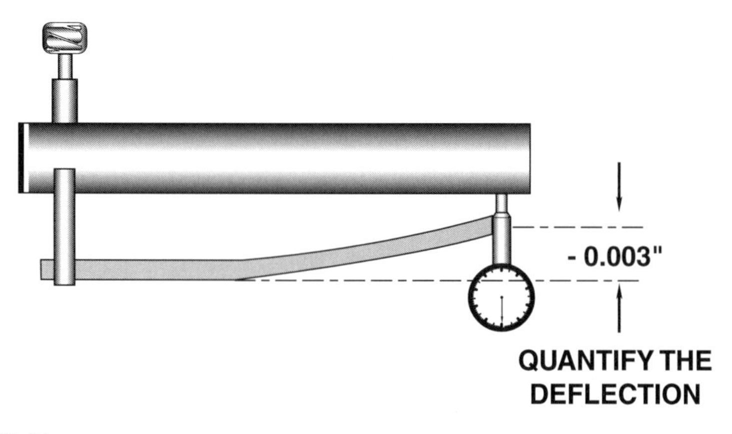

Figure 10-14

- This same procedure can be used during the actual alignment procedure to cancel bar deflection.

11. Perform a Preliminary Alignment

    - Bring the equipment shafts into a reasonable state of alignment with a machinist straight edge ruler and calibrated spacers before using the dial indicators. When the shafts are far out of alignment the dial indicators will make numerous revolutions causing confusion. It is much better to perform a preliminary alignment before applying the indicators.
    - Double check the distance between shafts with the recommendation of the coupling manufacturer.

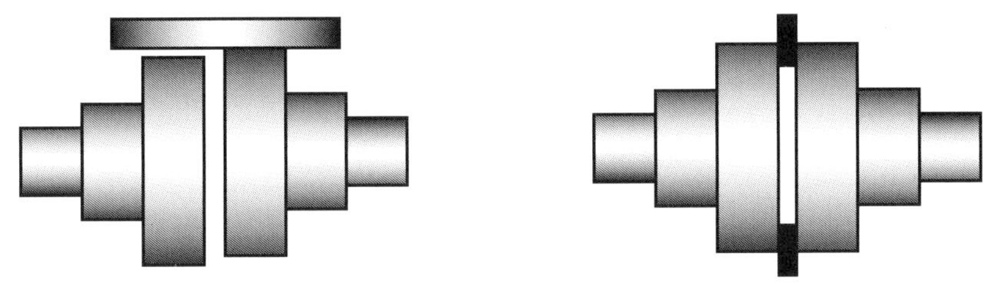

Figure 10-15

Equipment alignment sequence

Develop and practice the following alignment sequence.

Typical Steps

- Secure the pump to the base
- Always begin with the thinnest shims
- Use the minimum number of shims under any foot.

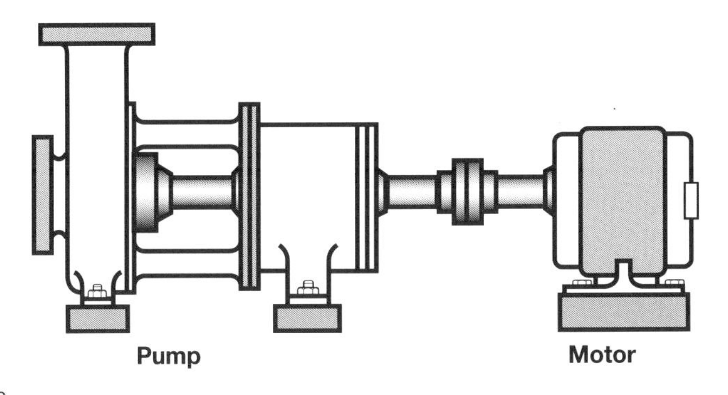

Figure 10-16

Coupling alignment

Don't use a flexible coupling to compensate for misalignment between the pump and motor shafts. The purpose of the flexible coupling is to compensate for temperature changes and to permit some axial movement of the shafts without interference, while they transfer energy from the motor to the pump.

There should be enough space between the coupling halves so that they don't touch should the motor shaft move forward toward the pump. This space should also consider movement due to wear in the pump thrust bearing. The coupling manufacturer specifies the minimum separation dimension between the coupling halves. You'll need a machinist rule and thickness gauge or feeler gauge to perform a rough alignment.

Before starting the alignment procedure, disconnect the coupling halves. First, verify the rough angular alignment inserting feeler gauges at four points (90°) around the faces between the halves. The alignment is correct when the feeler gauge distance is the same at all measured points.

The rough parallel alignment is done by placing the machinist rule across both coupling rim surfaces in the upper position, lower position, and both lateral points. The parallel rough alignment is correct when the straight edge rests uniformly on both rims at all four positions. Be sure that the coupling rims are concentric with the pump and motor shafts.

Don't start the pump until after completing all the previously mentioned points, and any other specification mentioned in the operation and maintenance manual of the pump provided by the pump supplier. Not doing this could cause equipment damage and even personal injury. It might even void the pump guarantee.

Bearings

Introduction

In order to understand bearings and their application in the world of pumps, it's best to consider some of the fundamentals and terminology of bearings. Pump bearings have two general classifications: sleeve bearings and rolling element bearings. The sleeve bearing is used mostly on reciprocating rods and shafts and on some low rpm rotating shafts. For rotating and centrifugal pump shafts the rolling element bearings have almost displaced the older sleeve type bearings. The precision rolling element bearing may have round balls, or rollers in the form of spindles, cones, or needle rollers as the rolling elements. The rolling elements move within an inner ring or race and an outer ring or race. With pumps, the inner ring mounts onto, centers, and rotates with the spinning shaft. The outer race is stationary and seated into a bore in the bearing housing. This arrangement may be reversed in some special duty pumps.

TYPE	DESCRIPTION
Ball Bearing – Conrad Deep Groove	This is a single row bearing. It can handle moderate radial and axial loads from any direction.
Ball Bearing – Heavy Duty	This is also a single row design, but it has more balls in the assembly than the Conrad bearing.
Ball Bearing – Double Row, not self-aligning.	These bearings have two rows of balls and can handle 50% more radial load than the Conrad type bearing. Alignment is critical.

TYPE	DESCRIPTION
Ball Bearing - Double Row, self-aligning	These bearings have two rows of balls and they can handle some shaft misalignment. The axial or thrust loading is limited.
Ball Bearing - Angular Contact	These are single row ball bearings capable of handling extreme axial or thrust loads in one single direction only. They are generally mounted in pairs to handle any opposite axial/thrust loads.
Roller Bearing - Cylindrical	This bearing is best suited to handle radial loads. It doesn't work well with axial/thrust loads. This bearing is popular in electric motors.
Roller Bearing - Conical	These bearings are good handling extreme axial / thrust loads in one single direction only.
Roller Bearing - Needle	These are a type of roller bearing with a roller diameter normally less than .250" and about a 1:6 diameter to length ratio, thus the term needle.
Roller Bearing - Spherical Symmetrical	These spherical rollers are barrel shaped and mounted in two rows rolling inside a common inner and outer race housing.

Each type of bearing is manufactured with different classifications of play or internal tolerance. This play or tolerance refers to the distance between the surface of the balls or rollers and the internal and external raceways with axial or radial movement. The bearing tolerance is rated with the codes C1, C2, CN, C3, C4 and C5. CN code is the normal, standard or average precision. C1 is the most precise tolerance (the tightest) bearing and C5 is the least precision tolerance (the loosest) bearing. A bearing without a stated code is normally a CN rating. Standard pump bearings are mostly C3 or higher, slightly forgiving and loose. With this understanding of pump bearings, you can see that it is not convenient to replace a C3 pump bearing with a C1 bearing. Because of its strict tolerance, a C1 bearing can handle less thermal expansion and lubricant contamination that occur with normal pump operation. A C1 bearing is less forgiving with thermal expansion and contamination and will fail prematurely.

As we've seen in previous chapters, the pump has basic head and flow specifications. The pump manufacturer installs bearings suited to the pump's design and general utilization. Based on this logistic, it's always best to use and replace the same type of bearing in a maintenance function.

Bearing lubrication

Oil lubrication

Oil is a popular pump bearing lubricating fluid. There are two classifications of lubricating oils: synthetic based and petroleum based. Synthetic lube oils were developed principally to improve their high temperature stability and overall temperature range. Petroleum based oils are lower in cost and have excellent lubricating properties. Animal and vegetable based lube oils are rarely considered for bearings because they deteriorate, spoil and form corrosive acids.

As a lubricating fluid, lube oil can be circulated, filtered, cooled, heated, pumped, atomized, etc. For these reasons it's more versatile than grease. It's convenient in severe applications involving extreme temperatures and high velocities. The oil level and flow through high-speed bearings is critical and these parameters should be monitored appropriately. Because oil is a liquid, it is more difficult to seal and contain inside the bearing chamber than grease.

Viscosity is the term used to indicate oil's thickness or consistency. The viscosity test is a measure of the fluid's film strength. Excessive viscosity can lead to the rolling elements skidding inside the bearing races. This leads to raceway damage and overheating of the oil. Inadequate oil viscosity can lead to premature bearing seizure from metal to metal contact. Other properties to consider in selecting and using oil are: corrosion protection additives, the pour point, the flash point, the viscosity index, carbon residue, and the neutralization number. Consider the loads imposed and operating temperatures of the bearings when selecting the oil and its viscosity. Correct selection is essential for long bearing life. The lubricant manufacturer publishes a data sheet detailing the characteristics of their products. You should get this sheet, study and understand the properties of the lube oil you're considering for use in your pump bearings.

Among oil's advantages are:

- Excellent friction reduction at high speeds.
- Provides good lubrication at very high or low temperatures.
- Good stability.
- Heat transfer coefficient is 0.5 BTU/pound/degree Fahrenheit.

- Works well through heat exchangers with refrigerated water for additional cooling in large high-speed bearings.

There are some disadvantages to using oil, including:

- Difficult to seal and retain inside the bearing chamber requiring frequent re-filling.
- Oil levels must be checked with more frequency as the temperature rises.
- Lower viscosity and more leaks as the temperature rises.

Grease

If you combine petroleum based or synthetic based oil with a suitable thickening agent, you get grease. The thickening agent can represent between 3% and 30% of the total volume of grease.

The amount of thickening agent and the oil's viscosity will determine the consistency or stiffness of the grease. For general industrial service, popular thickening agents used to make grease are soaps of calcium, sodium, aluminum, and lithium, and combinations of these. All greases have certain characteristics, but they should not be categorized for usage according to the thickening agent. For example, just because grease is thickened with sodium soap doesn't mean it is necessarily best in a particular service. No particular grease can satisfy all requirements, although considerable effort has been invested in developing a multi-purpose lubricant.

Grease is selected for use in a pump bearing for its advantages, including its resistance to dripping and running, and it's easy to seal and retain in the bearings. Grease also has some disadvantages. It is subject to separation and oxygen decomposition. It is difficult to clean and remove old grease from the bearing assembly.

Bearing lubrication in pumps

The majority of industrial pumps are designed with a horizontal shaft assembly. Therefore, more pump bearings are lubricated with oil contained in the bearing chamber. Grease is preferred for bearings on vertical pump shafts. Lip seals, labyrinth seals or mechanical seals are used to seal the bearing housing where the spinning shaft passes through, keeping any contamination out of the bearings and holding the lubricant into the housing. Seals are discussed later in this chapter.

Correct oil level is critical. Consult the owner's manual. As a general guideline, the oil level should be half way up the lowest ball or roller in the bearing assembly. Too much, or too little oil will lead to premature

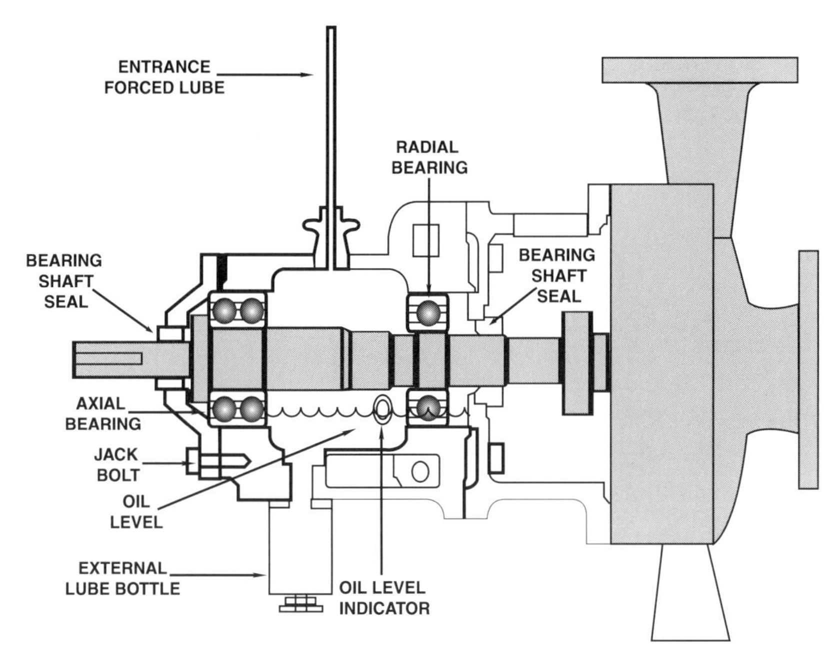

Figure 11–1

bearing failure, which leads to coupling and motor failure on the power end, and mechanical seal failure on the wet end.

Oil mist systems exist to provide continuous minute quantities of oil fog to the rolling assembly. These systems normally employ an additional pump, atomizer, and filter. These systems are gaining popularity in hot applications, or with heavy thrust and radial loading. The oil fog is sprayed into the bearing chamber with either a wet sump or a dry sump. The wet sump method provides the bearings with a bath (the liquid oil level) and a fog spray. See Figure 11–1.

The dry sump method of oil misting has no liquid oil contained in the bearing chamber. Instead, the entire chamber is filled with the atomized oil fog. See Figure 11–2, next page.

Both the dry sump and wet sump method of oil misting have a slight positive pressure in the bearing housing. This prevents contaminants and even humid air from entering the bearing chamber.

Bearing failure

Pump operators and pumping systems are plagued by unexpected premature bearing failures. Even if the cost of the bearing is small, the

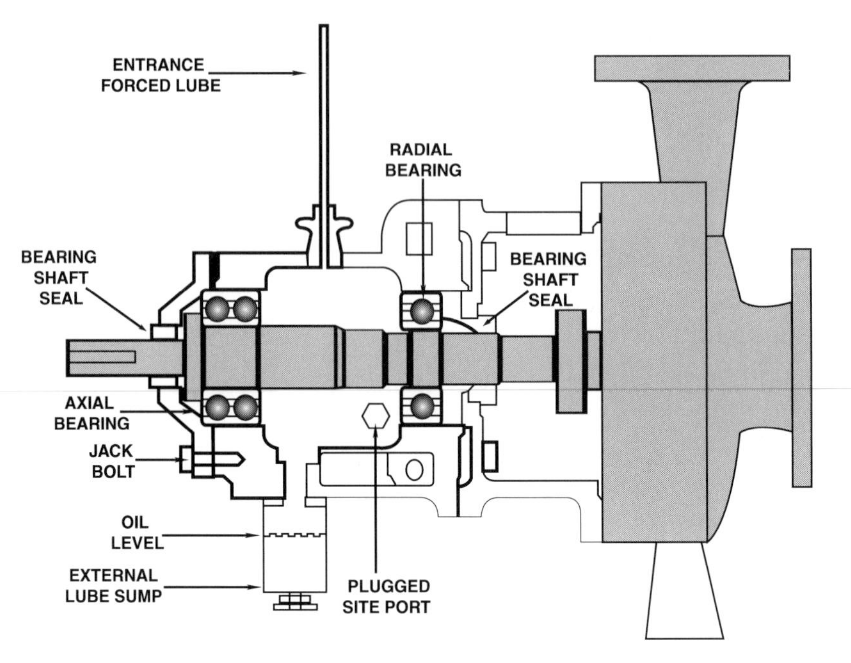

Figure 11-2

direct and indirect associated costs of its failure and replacement can be substantial. For example, a pump bearing may only cost $20.00 to buy, but its failure could also take out a mechanical seal. Now, besides the cost of the bearing and mechanical seal, is the cost of disassembly and reassembly of the pump. And, there will be other replacement parts to change although they may or may not have failed. Some of these would be the casing gaskets, pipe flange gaskets, set screws, snap rings, clip rings, wear bands, shims, oil seals, nuts and bolts, not to mention the oil or grease lost. Then there is the time dedicated to the repair, which is also the time lost from production.

Most often, the bearing or the lubricant is blamed for the failure. This is like blaming the fuse for an electrical failure. The failure is most likely a result of some abnormal operating condition, or lack of proper maintenance. In short, something causes the bearing to fail prematurely.

Among the most common causes of premature bearing failure are the following:

1. Improper mounting.

   Improper procedure when mounting the bearing on the shaft is one of the most common causes of premature failure. Roller bearings and ball bearings are precision devices, and correct installation practices are very important. They must be stored correctly and handled correctly to give good service life. The shaft and housing dimensions must be within limits specified by the bearing manufacturer. Shaft to motor alignment is also critical.

You should strictly follow correct and acceptable practices when removing the old bearing and installing the new one. Cleanliness is the order of the day. You'll need a clean work area, clean hands, and cleaning cloths without fuzz, lint, or strings. So much of premature bearing failure is the direct result of not observing these basic concepts.

2. Vibration Brinelling

   Maintenance people are not normally familiar with vibration brinelling, but this is also a common cause of failure. The brinell marks themselves are small, even invisible indentions in the bearing raceway. They result from vibrations or shocks originating outside the bearing. Common sources would be cavitation, bent shafts, unbalanced rotary assemblies; shock thrust loads, slapping v-belts, etc. These vibrations cause the balls and rollers to jam into the raceways causing the imperceptible indentations. The races eventually take on the appearance of corduroy cloth or a washboard effect. It's like driving a car at high speed over a rough roadway. The surfaces of the balls and rollers begin breaking away, thus destroying the bearing. All bearings coming out of service should be disassembled to examine the internal rotary and stationary surfaces.

3. Dirt and Abrasion

   Careless handling during storage and assembly damages a lot of bearings and lets dirt get in, thus leading to premature failure. Dirt, sand and dust contamination between the balls and races of a new bearing can start a round of 'false brinelling', ruining the bearing even before it goes into service. Dirty sweaty hands, damp cloths, humid air and even the morning dew can start a rusting process that will destroy a new bearing. Bearings should be kept clean. Some studies indicate that more than 90% of all bearing failure is attributable to abrasive dirt entering the bearing before and during its installation. A grain of dirt or sand trapped between the ball and race of a precision bearing has the same effect as running a race with a rock stuck in your shoe.

   Sleeve bearings on some older, slower, larger pump shafts can withstand dirt contamination better than ball and roller bearings. This is because the tolerances are not so strict with sleeve bearings, the surface area of contact is greater, and the lubricant flushing action is better. The sleeve bearing material of construction is normally softer than the shaft material, and abrasives can be absorbed into or imbedded into the bearing material without destroying it. This is certainly not the case with precision ball and roller bearings.

- Bearings should not be left exposed overnight. Coat bearings with clean oil and cover them with clean oil or wax paper.
- Clean the internal bearing housing and shaft seat before installing the bearing.
- Clean and paint all un-machined internal surfaces inside the bearing housing with oil resistant paint.
- Do not spin a new bearing by hand or with compressed air. This introduces dirt and grit and causes it to imbed into the protective grease.

4. Inadequate Lubrication

   Inadequate means not enough lubricant, and it means the wrong type of lubricant. If the lubrication is insufficient, the bearing suffers short life from too much friction, high heat, and metal-to-metal contact between rolling and stationary elements. For horizontal shafts, the proper oil level is: half way up the lowest ball or roller in the bearing assembly. Some pumps have a site level indicator showing the oil level inside the bearing chamber. It is recommended that you install these indicators if your pumps don't have them. Teach the operators and mechanics to pay attention to the indicators. As for greases, lithium based grease is for pumps. Polyurea based grease is for electric motors. If these two greases are mixed or confused, it will lead to premature pump bearing failure.

5. Excessive Lubrication

   Too much lubrication is as damaging to bearings as insufficient lubrication. It also leads to overheating. When the oil level is too high, the bearing balls and rollers crash into the oil pulling in air and bubbles leading to foaming. The foam and froth mixed with air cannot remove the heat.

AUTHOR'S NOTE

In cold climates, many homes and businesses use double and triple pane windows. The air pocket between double and triple pane windows serves to insulate so that heat is not lost through the windows. Likewise, air bubbles in oil prevent the heat from escaping.

The foam and froth in the bearing oil, increases the volume and artificially raises the oil level, which leaks through the seals. When enough has leaked to stop foaming, the air bubbles leave the oil resulting in inadequate oil levels. Too much friction heat and failure is the result.

Bearing maintenance

Cleaning bearings and relubrication

A lubricant, either oil or grease, should always be present in the bearings in small quantities. If not, the life of the bearing will be compromised by damage to the bearing surfaces. This damage can be avoided with proper cleaning and re-lubrication. The intervals for cleaning and re-lubing the bearings are generally long periods.

It's easy to see when a bearing needs oil. Check the oil site level indicator. It's different with grease. It's impossible to determine when a bearing needs more grease. This is because the grease in the bearing does not suddenly lose its lubricating properties. These properties are lost gradually over time. Previous operating experience (history) is a good guide to determine when to add more grease. The intervals depend on the grease properties, the size and design of the bearing, the operating speed, the temperature, and humidity.

In important process pumps, the grease in a bearing should be changed every 12 to 18 months. This will assure a reliable pump operation and service because time alone causes certain deterioration in the lubricating ability of grease.

The intervals for cleaning and re-lubricating bearings should be more frequent if water or moisture is able to enter into the bearing chamber. Bearings can become contaminated from rain, hose-downs, pumps located under dripping equipment, dew, fog and condensation. Entrance points could be through inadequate, worn or failed shaft seals, the breather cap, and the lube oil fill port. Be sure the new grease or oil is not contaminated.

Grease is normally injected through a port called a zerk (or zirk) fitting, or by removing the bearing end cover or housing cap. When injecting grease mechanically or hydraulically, remember to open the drain or expel port. The new grease will expel the old grease under pressure. Also remember to close the drain port afterward. The amount of grease to be added is a function of the housing size and design and the size of the bearing. The grease should completely impregnate the bearing and fill the housing about 25% full. Too much grease leads to overheating.

To remove old grease from the bearing internals and the housing:

- Remove as much as possible by hand.
- Flush the bearing and housing with warm kerosene.
- Following by a flush with mineral oil SAE 10 viscosity.

If the old grease is caked and hardened:

- Soak the bearing and housing in heated kerosene.
- Slowly rotate the bearing by hand.
- Rinse the bearing with clean kerosene or degreasing solvent.
- Again, rotate the bearing's outer race by hand while applying a modest axial and radial load to the balls and races.
- Soak and rinse again as necessary until the bearing rotates freely and smoothly.

Once all the old grease is removed from the housing and the bearing is cleaned, it should be thoroughly inspected for damage. If the bearing is not damaged, it can be repacked with new grease of the correct type and consistency, and reinstalled in the equipment or stored for future use. To store the bearing, wrap it completely in wax or oilpaper and place into a storage box.

The following table contains some simple 'Do's and Don'ts' for handling and working with bearings. Memorizing and practicing these suggestions will extend the service life of rolling element bearings.

	Do's	Don'ts
1.	Make sure all tools and surroundings are clean.	Don't use chipped or dirty tools.
2.	Use only clean, lint-free cloths with no strings to wipe bearings.	Don't use cotton waste or dirty cloths to wipe bearings.
3.	Use only clean flushing fluids and solvents.	Don't use leaded gasoline to rinse bearings. The chemical additives are harmful to your health.
4.	Don't handle bearings by hand. It's best to use clean cotton gloves.	Don't handle bearings with wet or dirty hands.
5.	Remove all outside dirt from the housing before exposing the bearings.	Don't work in a dirty surrounding.
6.	Make sure the internal bearing chamber is clean before replacing bearings.	Don't scratch or nick any bearing, housing, or shaft contact surface.

	Do's	Don'ts
7.	Place bearings on clean paper.	Don't expose bearings to rust or dirt.
8.	Keep bearings covered with oil or wax paper when not in use.	Don't spin un-cleaned bearings by hand.
9.	Protect disassembled bearings from dirt and rusting.	Don't spin un-cleaned bearings with a jet of compressed air.
10.	Treat new and used bearings with the same care. Use an induction heater for installation.	Don't install bearings with mallets, or hammers and wood blocks.

AUTHOR' NOTE

Working recently as a pump consultant, I was in a failure analysis meeting with a chief mechanic who had some 45 years experience. A preventive maintenance inspector came into the room and reported that he had to stop a pump because the bearing temperature was too high. The inspector was young. He'd been working for 3 years in the plant. We went out to the site and the chief mechanic placed his hand on the bearing chamber, announced that the temperature was normal, and ordered to start the pump motor again. I put my hand onto the bearing chamber and I could barely maintain contact because it burned. Consider this. The inspector reported that the bearings were hot. I thought the heat was critical. (I've been writing this book for two years.) To the chief mechanic, the temperature was normal.

We put a contact thermometer on the bearing housing. The thermometer indicated 152° F. During all these years, this chief mechanic's hands had seen a lot of temperature, abrasion, and abuse. His hands and touch were much more resistant to heat than mine, or the PM inspector's hands.

The moral is: If you don't have 45 years experience, go get the right gauge or instrument before making decisions.

Measurement of bearing temperature

We often have a tendency to place a hand onto a bearing housing to measure the bearing's operating temperature. If it feels cool or warm, we're confident that all is well inside the bearing chamber. If the housing is hot to the touch, we get worried about a potential failure and we spend time and effort to lower the temperature, hoping to gain a clear idea of what's actually happening inside the housing.

The fact is, that the human hand is not a good thermometer and it can give false temperature signals. In studies of human touch defining 'hot', hot varies somewhere between 120° and 130° F, depending on the individual. The human hand is worthless above this arbitrary point to estimate temperature.

Rolling element bearings lubricated with grease can operate safely in the 200° F range. In fact, the upper temperature limit of the grease is the real operating limit of the bearing. It's not the bearing metallurgy. The temperature at which the grease carbonizes is the bearing's operating limit. Bearings are perfectly safe at 160° F. This is actually good for a bearing because of the expected lubricant flow at this temperature.

It's obvious that all bearings will operate at some temperature above the surrounding environment, without additional cooling. The resulting temperature is composed of three factors. First, frictional heat is generated inside the bearings from contact between the rolling and stationary elements. Second, conductive heat is added to frictional heat. This is heat from the shaft, bringing the temperature of the pumped liquid, and also the radiated heat of surrounding equipment in the area. Third, the amount of heat to be dissipated away from the pump's bearings is a function of the conductivity of the lubricant, the surface area of the bearing housing, and the temperature, and motion of the surrounding air. These three factors work to bring about a stable operating temperature. This temperature should be less than the upper limit of the oil or grease.

You need to investigate an unexplained rise in the bearing temperature. This could indicate imminent failure. You can add one more shot of grease, but if the temperature doesn't reduce immediately, don't continue adding more grease. First, rule out obvious reasons for the increase. It could be that the temperature in the surrounding area has changed. Has the weather changed? Has new heat generating equipment been installed in the vicinity? Has there been a change in the temperature of the pumped liquid? Next, check the assembly for unnecessary thrust and radial loading, coupling misalignment, or over-tightened pump packing.

Remember again that the temperature can go up from improper lubrication practices. Excessive grease causes the bearing to sling and pack the grease against the internal housing wall. The grease becomes an insulator and the bearing will run dry if the grease cannot return to the sump. Excessive oil causes foaming and air bubble entrainment as the rollers and balls crash into the fluid. Air is a good insulator and the air doesn't lubricate or dissipate heat. Insufficient oil or grease leads to increased friction from metal to metal contact. Inadequate oil and grease are also sources of excessive frictional heat.

Pumps that handle hot fluids have bearing chambers designed with thermal jackets and heat exchangers installed at the factory. These devices have connections for isolated water flow through, or around the bearing housing. You should not use high temperature grease with artificially cooled bearings. The grease won't flow properly. The result will be consistent with inadequate lubrication.

Sometimes bearings seem to run hot at pump start-up. This may be heat actually generated by the bearing seals and not the bearings or inadequate lubrication practices. After the seals seat and settle, the temperature should go back to normal.

Bearing seals

The mechanical seal for bearings

Among the newer developments for industrial pumps is the bearing mechanical seal (Figure 11–3). These seals are designed to run in the same space provided for lip and labyrinth seals. There are two basic concepts to designing these seals. One concept incorporates rotary and stationary faces held together with spring tension like standard process pump seals. The other basic concept utilizes magnets to hold the faces together. The flexibly mounted faces permit a small degree of axial and radial movement of the shaft without compromising the sealing ability.

These seals perform well to completely separate the environment inside the bearing chamber from the environment outside the bearing chamber. These seals have proven effective in retaining grease and oil and especially the oil fogs (described earlier in this chapter) inside the bearings. By holding positive pressure, neither contaminants nor humidity can enter into the bearings. It is recommended to close and plug the breather cap, or to use this port to install humidity, temperature and level sensors to monitor the bearings.

Figure 11–3

The labyrinth seal

The word labyrinth means 'a tortured pathway'. In the Dark Ages, elaborate gardens, stone walls and an artificial lake with a drawbridge would lead up to the main gate of a King's Castle. This was the labyrinth. An attacking army would have to march through the garden, around the stone walls and swim the lake containing ferocious crocodiles, to attack the castle. All the while, the King's soldiers were shooting arrows at the attacking army as they marched back and forth and swam the croc-infested lake.

As a bearing shaft seal mounted into modern industrial pumps, the labyrinth seal is composed of a rotary unit that spins with the shaft and a stationary unit mounted into the bore of the bearing housing around the shaft. Labyrinth seals are considered 'non-contact' seals. The rotary and stationary units do not actually touch each other. However, they are in very close proximity. Its operating principal utilizes centrifugal canals or grooves with openings to an external gravity drainage.

The dual purpose of the labyrinth seal is to prevent external contaminants, like dust and water, from entering into the bearing housing, while it maintains the lubricating grease or oil inside the bearings. If a dust particle or drop of water tries to enter into the bearings through the seal, it is caught into the labyrinth of centrifugal spirals and ushered toward the external drainage. If the bearing lubricant tries to exit the housing through the seal, it is trapped into it's own labyrinth and returned toward the oil sump.

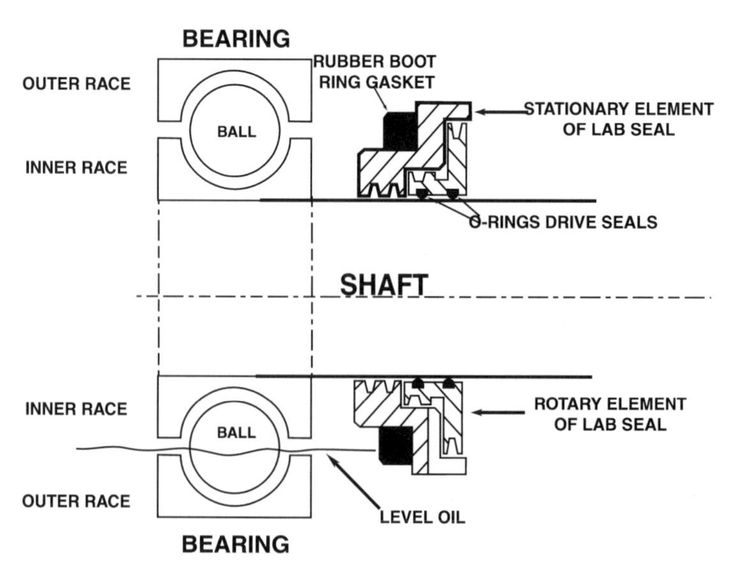

Figure 11-4

Most commercial models will ride into the same radial space provided for earlier lip seals. They may require some additional axial space on the pump shaft, but this normally doesn't interfere with other obstructions or equipment. Even if pump modification is required to accommodate the labyrinth seal, it is an improvement over the lip seal. Remember that the bearing housing was first bastardized to accommodate the lip seal. Any further modification to accommodate the labyrinth seal will not affect the service of the pump.

Labyrinth seals work best when the pump is running. Centrifugal force favors the labyrinth seal's action. Earlier models were only specified for horizontal pump shafts. Later models are designed for both horizontal and vertical pump shafts and effectively perform their function whether the pump is running or off.

The lip seal

The lip seal, or oil seal, used on modern centrifugal pumps is borrowed from the automotive industry. The lip seal was born with the invention of the automobile transmission and the universal joint in the early days of the family car. It would effectively retain the transmission fluid and U-joint grease on jalopies with rumble seats. It really hasn't changed much in design since the 1920s.

The outside diameter of the lip seal fits and seats into the housing bore (transmission or pump). The inside diameter, with the elastomeric lip, rides onto the spinning shaft (whether vehicular drive shaft or pump shaft).

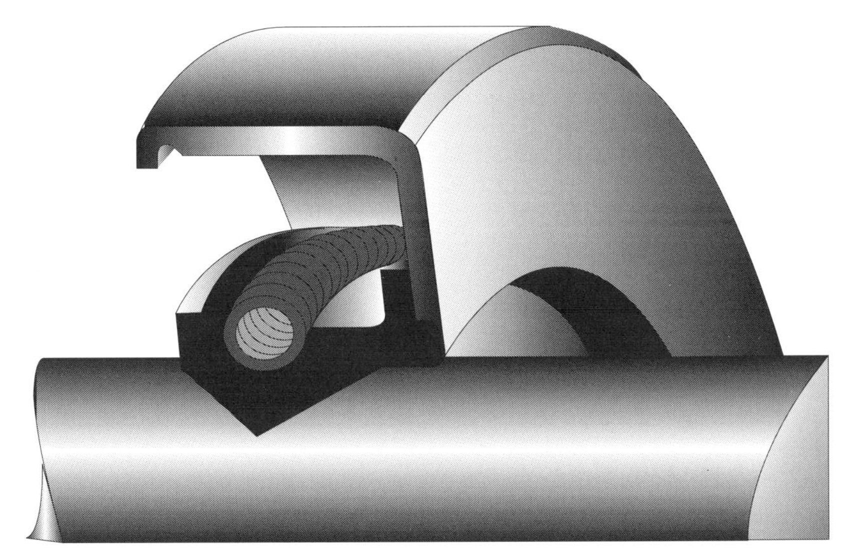

Figure 11-5

These seals work well within their designed life. Their designed life is about 2,000 hours. An automotive drive shaft spinning at about 1,800 rpm would move the car at approximately 50 miles per hour. 2,000 hours would be equivalent to about 100,000 miles on a car. 2,000 hours (at 1,800 rpm) on a pump would be equivalent to about 83 days at 24/7 operation. Many mechanics have questioned the logic of installing a 3-month seal to protect a 5-year bearing.

At about 2,000 hours, one of two things can happen to cause failure to a lip seal. Either the frictional heat from the spinning shaft burns and cooks the rubber lip, or the rubber lip eats a groove into the shaft. How can a soft elastomeric rubber lip cut a groove into a stainless steel shaft?

One of the components of stainless steel is chromium. A layer of chromium oxide is visible on the surface of stainless steels. That's why stainless steel appears to be chromed. As the stainless steel shaft spins under the rubber lip, the chromium oxide particles imbed into the rubber lip.

AUTHOR'S NOTE

Chromium Oxide is present in just about every maintenance shop in the world. We call it the GRINDING WHEEL. The abrasive material in your electric grinding wheel is Chromium Oxide. Cheap wheels may tend to use aluminum oxide.

After a few revolutions, the rubber lip of the oil seal becomes an abrasive lip, which eats a groove into the stainless steel pump shaft. The rubbing action abrades the pump shaft, removing metal, and depleting the chromium content of the stainless steel, which further accelerates its erosion.

When the rubber lip can no longer maintain contact with the spinning shaft, the oil or grease can leak out of the bearing housing. Contaminants can enter into and destroy the bearings. When the lip seal is changed with the bearing change, the new lip rides into the old groove cut by the previous lip. That's how a $6-dollar rubber lip seal can take out a $300.00 bearing and an $800.00 stainless steel pump shaft ... about 4 times per year.

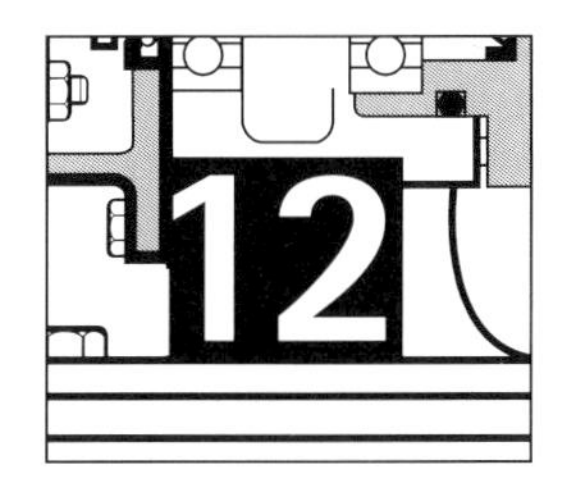

Pump Shaft Packing

History

In the beginning of recorded time, primitive man began building boats for fishing and to explore his world. The rudder appeared in some of the original designs of boats. The rudder was a specialized type of oar. It was composed of a handle on the upper end, and a shaft mostly mounted in a vertical fashion. The shaft passed through a hole in the bottom of the boat. Often the hole was below the water line. The lower end of the rudder shaft was submerged into the water. The lower end of the shaft was designed with a flat palette or paddle. This flat paddle was called the 'tiller'. The sailor up in the boat could rotate the rudder and thus steer or navigate the boat with the tiller in the water below.

The hole in the bottom of the boat, where the rudder shaft passed through, was a point of leakage where water would enter into the boat. So the early boat builders had to design a method of preventing the entrance of water. They designed a box-type housing around the hole with a circular gland type press. The sailors would stuff or pack their old clothes, hair, rotten ropes, old sails and leather scraps into the box-type housing. The word 'stuffing box' was born. The purpose of the circular gland type press was to squeeze and compress the stuffing, called 'stopa' (pronounced STOH-pah), into the box, creating a seal between the rudder shaft and the hole in the bottom of the boat. This prevented the entrance of water into the boat. The term 'stuffing box' is still used today referring to pump design. (In Spanish the word for stuffing box is 'prensaestopa' or literally 'stopa press'.)

Vegetable fibers

Aboard every sailing ship, there was a sail maker/tailor. This tailor's job was to make and repair flags, seal holes in the sails, in the sailor's clothes, and the hole in the bottom of the boat. The tailor would fashion stuffing box 'stopa' material from saved scraps of clothing, old sails and ropes. Out on the high seas, whenever a boat came upon an island, the sailors would disembark to search for food, fresh water, and stopa material. With some luck, the sailors found wild plants of cotton, jute, ramey, linen and hemp. Without luck the sailors returned to the boat with vines, root sprigs, and tree bark. They saved strings from mango seeds, corn shucks, and even the feathers, hair and hides of the animals that they hunted for food. The tailor would take these materials and form threads for sewing. The tailor would weave the threads into patches for the sails. Some threads would be formed into strings and ropes for hanging sails. Other threads and strings would be formed into stopa to seal the rudder shaft. In the port cities, the ship supply agents began selling prepared stopa, formed with linen and cotton lubricated with animal fat and wax, ready to stuff and press into the stuffing box around the rudder shaft. This stopa was resistant to the abrasive rudder shaft and the salty seawater.

Out on the high seas, the sailors would tighten the stopa around the rudder shaft and the friction would hold the tiller steady pointing the boat toward the horizon or a distant star. At times of war, or upon arriving into a port and dock, the sailors would loosen the stopa gland to easily navigate the boat. With the loosened gland, the seawater would enter into the bilge. An apprentice sailor would get a bucket and begin bailing the bilge, hauling the water overboard.

Reciprocating action

The ancient sailors would slowly rotate the rudder shaft to navigate the boat. This ancient design, the rotating rudder shaft and the stuffing box, has continued in existence down through the ages to today from the beginning of recorded time. The moment arrived when the rotary action was replaced with reciprocating action. In 1712, the reciprocating steam engine became a reality. A century later, after numerous failures, the steamboat was presented to a waiting public, able to navigate upstream against the current in rivers. Inside the engine, a load of steam was discharged against a piston and reciprocating shaft. Through a camshaft mechanism, the reciprocating shaft made propulsion paddlewheels rotate. In order to contain the steam inside the cylinders with the reciprocating rods and pistons, the old stuffing box design was incorporated, with its box housing, gland, and stopa material.

Packing

The new increased demands on the stuffing box and stopa of the steam engine are obvious. The old rudder shaft of the ancient boat only moved sufficiently to change the direction of the boat. The reciprocating shaft of the steam engine is in constant movement, with more velocity and friction. Compare the temperature of seawater with the temperature of steam. On a sailboat rudder, the stopa had very little pressure to hold back (2.31 feet of depth is 1 psi). With refinements and improvements in steam engines, the pressures rapidly climbed through 10, 30, 50, 100 and 200 psi.

The industry stopped using the word 'stopa', and adopted the word 'packing'. The new packing stuffed into the stuffing boxes on reciprocating steam rods could withstand the temperatures, abrasion, and pressures generated by steam. Asbestos, which comes from mines in rocks and mineral fibers, became a popular component of braided packing for high temperature applications. New lubricants, mineral and petroleum based, could survive the frictions and temperatures present with the constantly and rapidly moving shafts. Packing construction, braided tightly like a square rope, with surfaces designed to seal against the shaft, and the stuffing box wall, could contain the higher steam pressures.

Shortly after the development of the reciprocating steam engine, the positive displacement pump was born. These pumps could seal and generate pressures but with one weakness. The flow, or quantity of fluid that could pass through the pump, is a function of two factors: first, the size of the pump casing, and second, the motor's speed. The reciprocating steam engine is powerful by design, but slow. With the existing steam engines, it was necessary to increase the size of the pump in order to pump more. The reciprocating pump is only able to capture, move, and expel a fixed quantity of fluid according to the size of the casing. Fabricating large pumps brings its own problems of raw material, the mold construction, the heating and melting of the iron, the weight, transportation and maintenance.

Rotary action

Reciprocating action in engines and pumps was again converted back again into rotary action at the beginning of the last century. First, the rotary turbine was perfected. Shortly afterward, the internal combustion engine appeared. In the marine industry, the propulsion paddlewheels evolved into propellers. Ship design was greatly simplified with a direct drive shaft from the motor to the propellers. The weight and space of the complicated camshaft mechanisms and gears were eliminated. On land, the electric motor opened the door to the first practical centrifugal pump.

The practical centrifugal pump was a significant invention. The rotary motor (whether steam turbine, internal combustion or electric) resolved the issue of velocity. The rotary motor permitted the pumping of tremendous quantities of fluid with relatively small pump housing, compared to its reciprocating sister. The pumping action begins again with each revolution of the spinning impeller, instead of each slow cycle of the piston and rod.

AUTHOR'S NOTE

There is another important difference between the centrifugal pump and its reciprocating sister although it is not considered a weakness. By design, the flow of the reciprocating pump has to pulse, similar to the beating of the human heart. The centrifugal pump with the rotary motor and impeller has a constant flow without pulsing. With advances in technology, rotary positive displacement pumps have been developed, and when mated to electric motors can give a constant flow without pulsing.

Synthetic fibers

Since the beginning of time, water, either fresh or saltwater, was the only liquid to be pumped, and sealed with stopa or packing in large quantities. Other liquids like milk, beer, paints, oil, gasoline, solvents, tanning acid and medications were made in bottles, jars and barrels and carried from one place to another. With the capacity to move large quantities of fluid with motors and small centrifugal pumps, our forefathers recognized the advantages of expanding production of other liquids that were not water. And as pumped liquids expanded, the components of packing materials also expanded to effectively seal these liquids.

Throughout recorded time and history, the basic design for the stuffing box and the gland press has not changed, because it has to work to accommodate, seat, and squeeze the packing. The stuffing box existed in biblical times, the era of the Vikings, and the centuries of the Spanish and English explorers, sealing seawater and permitting the progress of man and history.

Stopa made the transition from sealing the rudder shaft in the bottom of a sailing vessel, to become braided packing that could resist the temperatures and pressures or steam and the high velocity shafts.

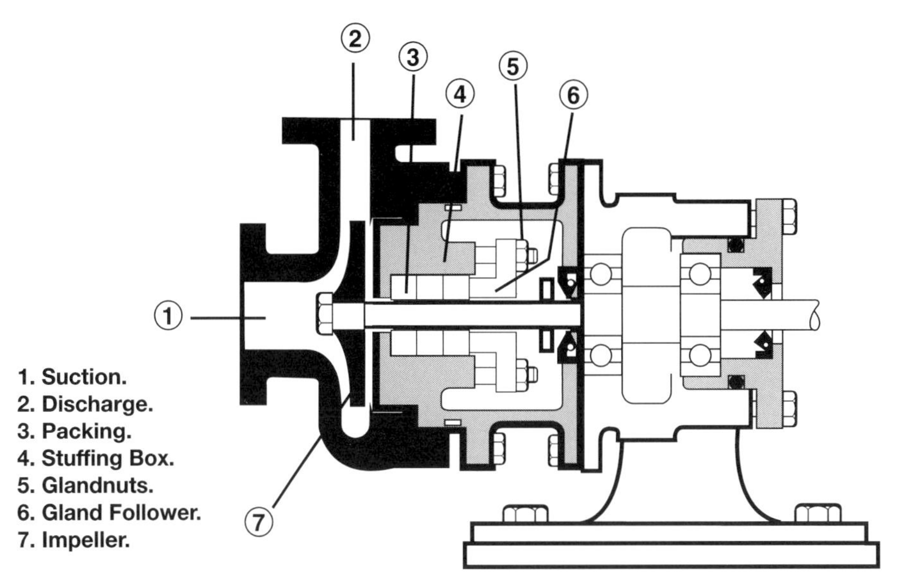

Figure 12–1 Typical back pull – out pump with packing

Likewise, modern braided packing can resist the temperatures chemicals, shaft velocities, pressures, and operational conditions that are found in today's industry.

Compression packing

Compression packing is frequently called soft packing because of the materials of construction used, cut to the proper length to form packing rings. These rings are packed into the stuffing box around the shaft. Through the stuffing box and the gland follower, the packing is compressed to accommodate the interior bore of the stuffing box and the shaft surface, sealing the pumped liquid.

In the case of valve stems where there is little or no shaft movement, the packing can be tightened and compressed to seal without any leakage. In the case of rapid rotational movement, as with the shafts of centrifugal pumps, or cycling movement in the case of reciprocating rods, the packing should experience some leakage for lubrication and cooling.

Packing comes in many forms, sizes, and materials of construction in order to meet all the needs of industry. And there are specific styles of packings to satisfy specific needs. The most common fibers used in the construction of packings include: Asbestos, Linen, Ramey, Jute, Cotton, Paper, Wool, Hair, Nylon, Rayon, Teflon, Fiberglass, Carbon

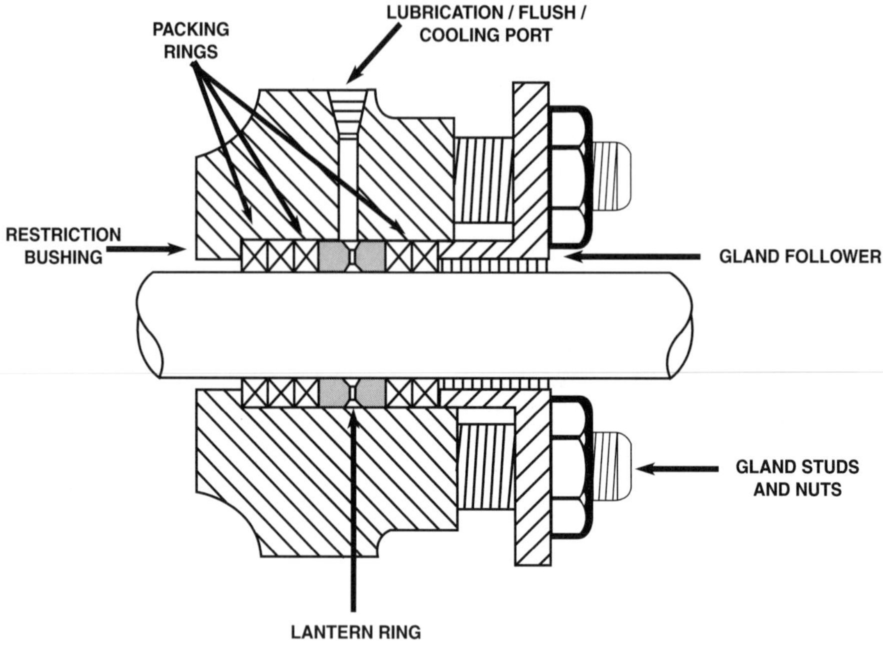

Figure 12-2

and Aramid. Some packings have metallic components such as: Lead, Copper, Aluminum, Iron, Stainless Steel, Nickel, Inconel and Zinc. Common lubricating components include: Animal Fat, Petroleum Oils and Greases, Mica, Graphite and Synthetic Lubricants. Still other packings have sizable quantities of rubber and leather. A general understanding of these materials of construction, their strengths and weaknesses, along with the equipment where they will be used, will reduce unexpected failures in the equipment, and as such reduce the equipment downtime and maintenance costs.

Pump packings should be selected for their ability to compress. When gland pressure is applied, the packing should react quickly to detain the leakage and not damage (score) the shaft or sleeve.

Another important consideration, often disregarded at the moment of buying the packing, is its effect on the shaft or sleeve. This is the most important factor to consider in selecting the appropriate packing. The cost of the shaft or sleeve will greatly exceed the few dollars saved by purchasing economical packing.

The lantern ring

The lantern ring performs three basic functions:

1. To supply new lubricant to the packing rings, which is normally lost in operation.
2. To supply 'back pressure', which aids in impeding the entrance of abrasive and corrosive material into the stuffing box. Abrasives and corrosives will damage the shaft or sleeve, and disintegrate the packing.
3. To cool the packing and shaft and dissipate the heat generated between these members, which will increase the normal service life of the packing and pump.

Oil, water, grease, or any liquid or substance compatible with the fluid are forced under pressure into the packing through the lantern ring by means of a connection on the stuffing box wall to provide these three functions.

Recommended instructions for packing a pump are:

1. On installing Teflon packings, place and seat each ring ringer tight. Do not use wrenches or pliers.
2. Bring the gland up to the packing rings and adjust the gland nuts by hand. (No tools yet.)
   a. Open the flush line to the packing.
   b. Start the pump.
3. Permit the pump and packing to leak generously for about 15 minutes to allow the packings to absorb the pumped fluid, swell, seat and adjust to the stuffing box wall and the shaft.
4. If excessive leakage continues, tighten the gland nuts $\frac{1}{6}$th of a turn with a wrench. Don't permit the temperature to rise in the packing.
5. Continue adjusting the gland nuts $\frac{1}{6}$th of a turn every 15 minutes until the leakage is controlled to about 1 drop per second per inch of shaft diameter.

PRECAUTION
Too many gland adjustments will cause the packing to crystallize and burn, which will shorten the packings' useful life and damage the shaft or pump sleeve.

The packing lubricant

The use of the appropriate lubricant is an important consideration. Generally Graphite, moly grease, and oils are good lubricants depending on the application. Many pumps are set-up to use cold water flushed into the stuffing box as a coolant, and lubrication for the packings.

Graphite is a common lubricant for general service. Molybdenum disulfide (moly grease) is an excellent lubricant because it creates a lubricating film barrier between the shaft and the packings, protecting both. It also works well with high temperatures. Mica is also a good lubricant for very high temperatures. However, it tends to contaminate the pumped liquid with particles. It certainly wouldn't be indicated in a milk pump. Any oil compatible with the pumped liquid is adequate as a general lubricant.

Be aware that the packing is destined to fail the moment it is installed. It has to resist every basic operating tendency of the pump. The packing consumes energy decreasing the pump's efficiency. It generates frictional heat by grabbing and abrading the shaft. The packing itself suffers from abrasion and corrosion originating in the liquid moving through the pump.

Repacking the pump is a difficult and dirty job and falls down to the bottom of the mechanics favorite 'to do' list. Extreme care should be taken to assure a proper installation. It's a job usually put off until the very last moment. Repacking a pump correctly takes a lot of time. Not following the correct repacking and start-up procedures results in short packing life.

Stages in the life of packing

The packing rings must totally fill and occupy all free space inside the stuffing box. The rings are composed of active and passive fibers (the passive fibers act as carriers to string and braid the active fibers). The packing also is impregnated with internal lubricants and bathed in surface lubricants. When the gland is tightened, the packing begins compressing and heating. The lubricants will eventually extrude out of the packings and the passive fibers will normally burn and carbonize into ashes. As the packing wears, it loses volume. There are generally four stages in the life of a packing ring:

1. Stage one is when the packing rings occupy the full space inside the stuffing box.

2. Stage two is when the packing gland has been tightened and the packing compresses. The space occupied in the pump is reduced.
3. Stage three is when the lubricant has extruded from and left the packing. At this point the packing should be changed
4. Stage four is when there is no lubricant and the carrier fibers have disintegrated into ashes. The packing becomes hard from calcification and destroys the shaft or sleeve.

Stage four packing should never be in the pump. It should be changed at stage three. When the gland follower butts against the stuffing box mouth and the gland nuts can no longer be tightened, then you've reached stage four. Stay away from stage four in the life of your packing rings.

Mechanical Seals

Pump packing

Many industrial pumps around the world still use ring packing as the shaft seal. These packings require continual maintenance. They need to be tightened and adjusted and it is necessary to change them frequently. Packings wear and damage the shaft or sleeve of the pump, which necessitates their frequent replacement. The leakage from the packing gland corrodes the pump, base plate and foundation. The drips can enter into the bearings and contaminate the bearing lubricant. This sacrifices the life of the bearings.

The majority of existing pumps that were originally designed for packings can be converted to use mechanical seals. Consider the following. We have many pumps in our private everyday lives. Practically all are adapted to mechanical seals. The family car has about 6 pumps with mechanical seals:

- The radiator water pump.
- The fuel pump.
- The oil pump.
- The power steering pump.
- The pump that sprays and cleans the windshield.
- The air conditioning compressor.

Do you have two cars? Then you have about 12 pumps with seals. And in your house, how many sealed pumps do you have? In the kitchen, mechanical seals are on the pump in the dishwasher, the blender, the refrigerator compressor, and the garbage disposal under the sink. Do you have a clothes washer? There are two seals on the clothes washer, one on the pump that fills and drains the washer, and another seal on the agitator shaft. How many drips are on the laundry room floor?

AUTHOR'S NOTE

Does your house have air conditioning, or a heat pump? Do you have solar heating, a Jacuzzi bath, swimming pool, aquarium, solar heating, a roof mounted water storage tank, a well, a power assisted commode, a motorboat, a jet ski, ATV or camper? Then you have even more mechanical sealed pumps. We have mechanical sealed pumps in practically all facets of our everyday life, and these pump seals last for years without problems.

The environmental laws favor mechanical seals over packings. The need to conserve energy favors mechanical seals. The needs to reduce labor costs and consumption of natural resources favor the mechanical seal over packings.

The majority of pump manufacturers offer their products with standard or optional mechanical seals. The mechanical seal manufacturers make seal models designed to substitute packings. The majority of pumps can be converted to mechanical seals without machining or design change. And still other pumps can be converted to mechanical seals with a slight design adjustment that doesn't affect flow or head. The conversion to a mechanical seal improves the pump's efficiency. The cost of the seal and the labor to convert the pump will be returned in reduced operating costs in just a few months.

The mechanical seal runs in the same space previously occupied by the packing rings (Figure 13–1 and Figure 13–2, next page).

The mechanical seal on the radiator water pump of your car has to work under severe conditions. This seal must resist the pressures and temperatures, corresponding to the velocities of the motor, and the variable operating times. This seal is not a precision seal (it has stamped parts rather than machined components) and the pump is a portable pump. The pump doesn't use a direct coupling but a v-belt pulley with radial loading. The seal must resist many vibrations commencing with the v-belt slapping and whipping.

The seal must also resist the vibrations from the explosions of internal combustion in the engine, chassis and wheel vibrations, and even potholes in the road. This seal must resist strong chemicals (anti-freeze, anti-rust agents, radiator stop-leak and sealant chemicals, gasoline and lubricant residuals), and also solid particles (rust, iron slag, minerals, asbestos fibers, and silica from the engine casting mold). In spite of all this, the mechanical seal on the water pump of your car can run 7, 10, even 15 years without problems.

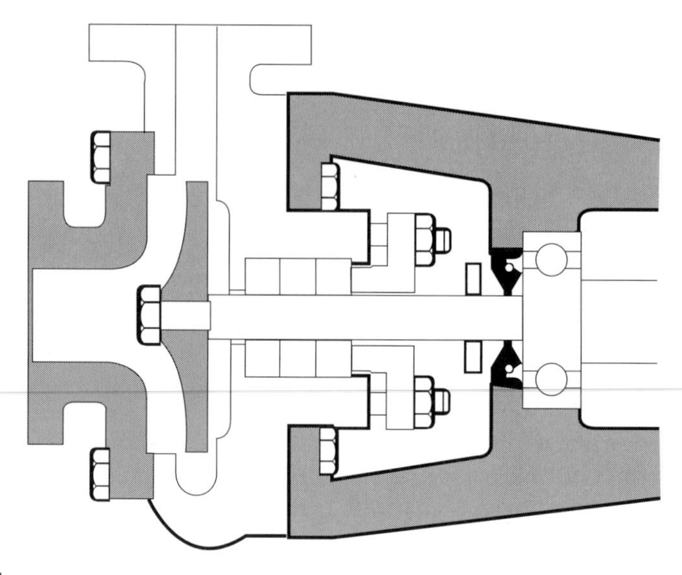

Figure 13-1

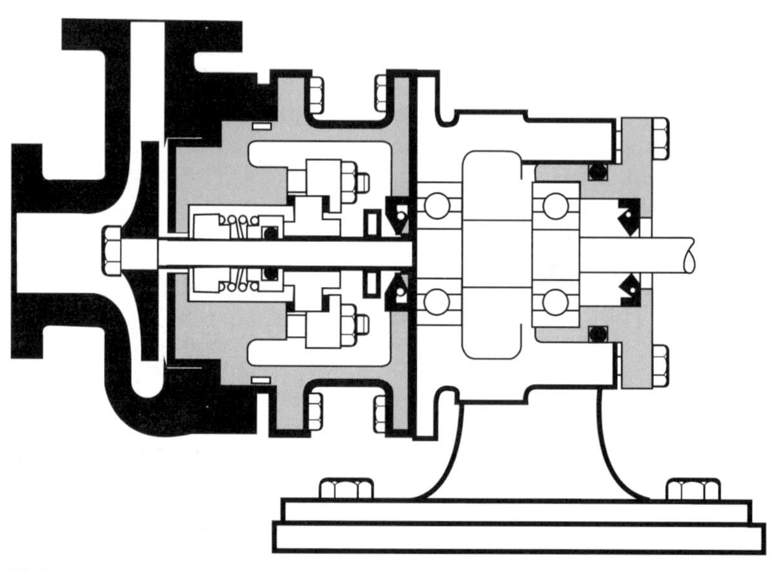

Figure 13-2

History

The mechanical seal was developed at the beginning of the last century. Its development coincided with the invention of the steam turbine, the dynamo and the electric motor.

In 1919, there was a patent on a mechanical seal showing spring mounted faces to compensate some axial shaft movement. The balanced mechanical seal was patented in 1933. The balance feature raised the pressures that the seal could withstand, and reduced the heat generated between the faces. Mechanical seals were used on the propeller shafts of submarines during the Second World War. Mechanical seals replaced packings on the water pumps of jeeps and passenger cars, and on refrigeration compressors in the 1940s. Oil refineries began specifying mechanical seals as standard equipment in the early 1950s.

The development of the mechanical seal advanced in parallel with elastomer technology. Mechanical seals using o-rings and other elastomer forms, benefited with improved shelf-life, maximum and minimum temperature limits, better chemical resistance, and higher pressure ratings.

To this day, balanced mechanical seals using o-rings are the standard in industry. Mechanical seals continue to evolve in sealing face technology, computer design, finite element analysis, cartridge designs, split seals, double or dual seals, and dry gas seals.

Nowadays, there doesn't really exist a liquid, condition, or pump operating situation that cannot be sealed successfully with a mechanical seal. With the help of mechanical seals, man has been able to explore the extreme pressures of the ocean depths, and the extreme vacuum of outer space.

AUTHOR'S NOTE

We say this because there are still people in industry who consider the mechanical seal to be something new. They say they'll continue to use packing in their pumps until mechanical seals are perfected. This is like saying they'll continue to use candles until the electric light bulb is perfected.

Mechanical seal manufacturers design their products to last 5 to 10 years. The majority of seals are designed for 40,000 hours of operating life. There are 8,760 hours in a year. This is approximately 5 years, running a pump at 24 hours per day, and 15 years for a pump running at 8 hours per day.

It doesn't matter who manufactures the seal, almost all seals have the same component parts, because all parts have to perform the same functions. The common parts in all seals are: the gland, the stationary face, the rotary face, the secondary seals, the spring, and the fastener to the shaft. Let's look at how mechanical seals are designed (Figure 13–3, next page).

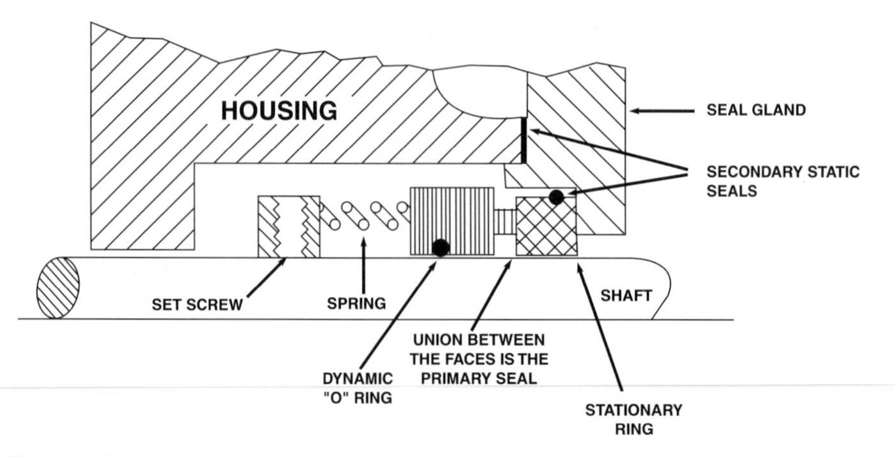

Figure 13–3

The mechanical seal

The mechanical seal is a device that forms a barrier between rotary and stationary parts in the pump. The seal must block leakage at three points (Figure 13–4):

- Between the faces (rotary and stationary) of the seal.
- Between the stationary element and the seal chamber housing of the pump.
- Between the rotary element and the shaft or sleeve of the pump.

These basic components and functions are common to all seals. The form, style, and design vary depending on the service and the manufacturer. The basic theory of its function and purpose nevertheless, remains the same.

The set screw that transmits the torque from the shaft is connected to the rotary face through the spring. It also provides for the positive and correct positioning of all rotary parts.

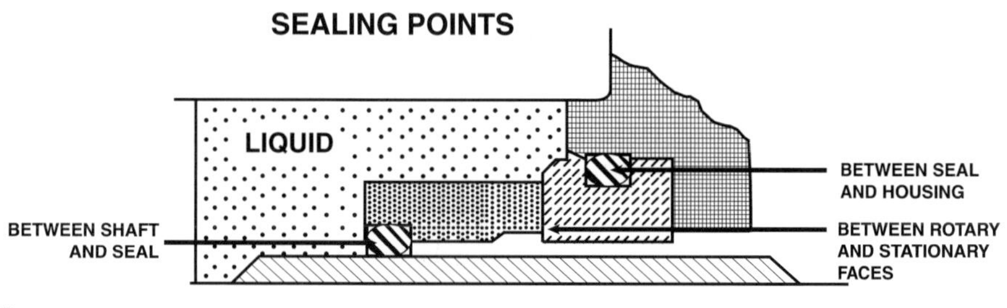

Figure 13–4

As the faces wear, the spring extends maintaining the rotary face in contact with the stationary face. The shaft o-ring should be free to move axially on the shaft within the operational tolerances of the bearings. This is called axial play.

The liquid's pressure in the seal chamber holds the faces together and also provides a thin film of lubrication between the faces. This lubricant is the pumped product. The faces, selected for their low frictional characteristics, are the only parts of the seal in relative motion. Other parts would be in relative motion if the equipment is misaligned or with loose tolerance in the bearings.

The single, unbalanced, inside mounted mechanical seal

This type of seal mounts onto the shaft or sleeve inside the seal chamber and pump. The pumped liquid comes into contact with all parts of the seal and approaches the outside diameter of the internally mounted faces keeping them lubricated. The environment outside the pump approaches the ID of the seal faces.

The pressure inside the pump acts upon the faces to keep them together and sealing up to about 200 psig. This is the most popular type of mechanical seal in clean (no solid particles or crystals) liquid service. A discharge bypass line connected to the seal chamber can provide additional cooling. Some people prefer a suction bypass connection with low vapor head (Figure 13–5).

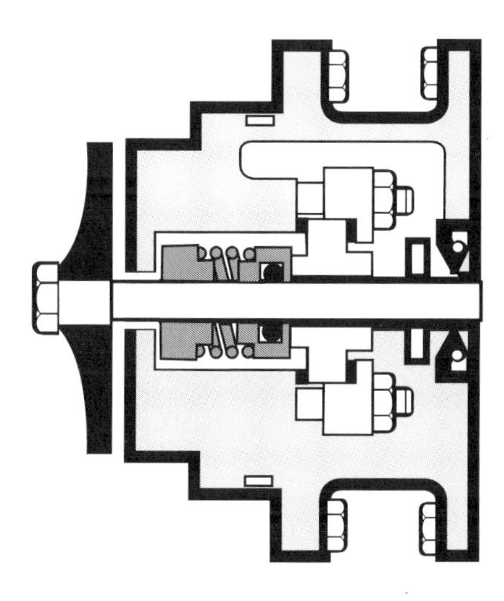

Figure 13–5

The single, outside-mounted, unbalanced seal

This type of seal has the rotary component and face mounted outside the seal chamber. The springs and drive elements are outside the pumped liquid. This reduces the problems associated with corrosion and the accumulation of pumped product clogging the springs. This seal is popular in the food processing industry. The pumped liquid arrives to the inside diameter of the faces and seals toward the outside diameter. The environment outside the pump approaches the OD of the face union. Pressures are limited to about 35 psig. Sometimes this seal can be mounted either inside or outside the pump. This seal is easy to install, adjust, and maintain. It permits easy access and cleaning of the pump internal parts, often required in the food processing industry.

The single, balanced, internal mechanical seal

This balanced seal varies the face loading according to the pressure within the pump. This extends the pressure limits of the seal (Figure 13–6).

The earlier balanced seals incorporate stepped faces mounted onto a stepped sleeve or shaft. Later models offer the balance effect without stepping arrangements. The pumped liquid approaches the OD of the seal's faces with atmospheric pressure at the ID of the faces. These seals are good in the 500 to 600 psig ranges, and they generate less heat than their unbalanced versions. They're popular in petroleum refining, and in general industry where some liquids are prone to easy vaporization.

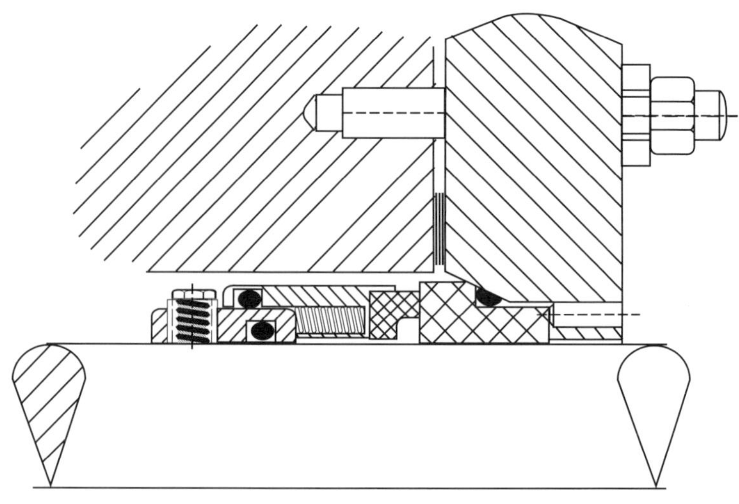

Figure 13–6

The single, balanced, external mechanical seal

Outside balanced seals permit sealing pressures up to about 150 psig. They offer the other benefits of outside seals, which make them popular in the food processing industry (Figure 13–7).

In the food production and OTC (Over the Counter) Drug industries, like milk, soups, cough syrup, and juices, outside balanced seals are quite popular. Their design permits easy cleaning of the equipment without pump disassembly. These seals are prominent in the chemical processing industry because all metal components in the seal are located outside the fluid. This avoids problems of galvanic corrosion.

Advantages of O-Rings

The majority of mechanical seal manufacturers make models that incorporate o-rings as secondary seals. These o-rings offer advantages over other forms of secondary elastomers.

1. The O-ring can flex and roll – There is no need for the shaft to slide and rub under the o-ring while under an axial load.
2. Availability – It's the only elastomeric seal form easily available in most cities and towns.
3. Material of construction – Almost all elastomeric compounds are available in the o-ring configuration.

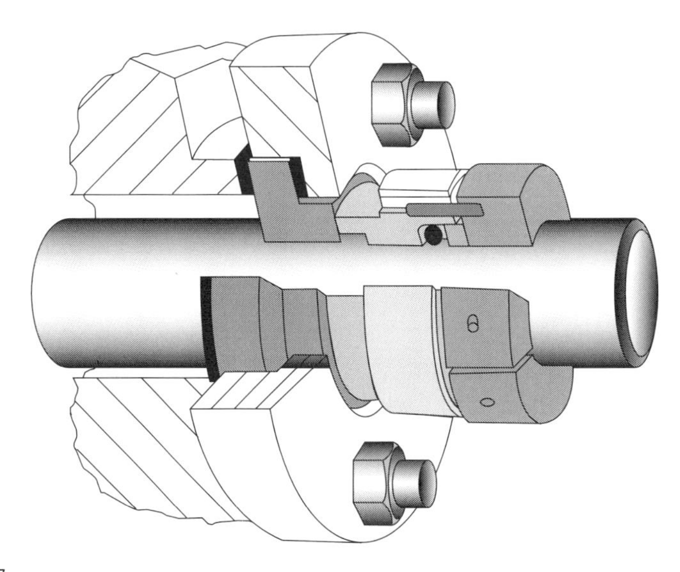

Figure 13–7

4. New materials – As soon as a new elastomer compound is developed, it becomes available and produced in the o-ring form. Examples are 'Kalraz', and 'Zalak' produced by Dupont, 'Aflas' produced by 3M, and 'Parafluor' by Parker Hannifin.
5. Reliability – Tolerances to .003 inch assure reliability.
6. Vacuum Service – The o-ring is the only common popular elastomeric form that seals in both directions.
7. High Pressure – Installed with back-up rings, o-rings are the standard in high pressure mechanical sealing.
8. Controlled Loading – The grip onto the shaft is determined by the seal design and the machined groove, and not the mechanic's ability to install the seal at the proper dimension and spring loading.
9. Easy Installation – O-rings slide easily over keyway grooves, impeller threading and stepped shoulders on sleeves.
10. Less Shaft Wear – Of course o-rings can damage and fret a shaft if the equipment is misaligned, but it takes much longer than with other designs.
11. Low Cost – Why pay more?
12. Misalignment – O-rings can compensate for some misalignment in the seal chamber face and bore better than any other elastomer configuration.
13. Impossible to Install Backward.

AUTHOR'S NOTE

These previous statements refer to designs incorporating o-rings mounted into machined grooves. O-rings should never be butted against a spring or charged with a spring load. They lose many of their desirable properties.

The balance effect

Another concept that has become quite popular in industry is the balanced mechanical seal. Most manufacturers offer seal models incorporating the balance feature. This balance is not a dynamic balance, but instead a relationship between the forces tending to open the faces in a mechanical seal and the forces tending to close the seal faces (Figure 13–8).

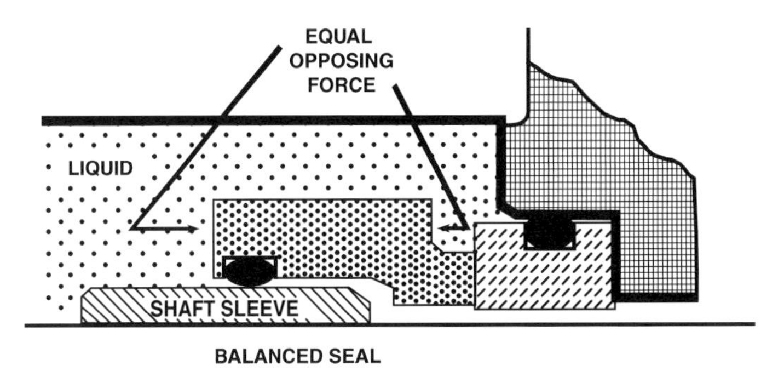

Figure 13-8

Advantages of balance

1. Less heat Generated – Less force between the faces indicates less heat generated. Heat is the principal reason for premature seal failure for two reasons:
   - It affects the elastomer, which is the seal component most sensible to temperature.
   - Heat can cause a phase change in the pumped liquid.
     - Some liquids like caustic soda can crystallize with additional heat.
     - Some liquids can solidify with heat (sugars).
     - Some liquids can vaporize with heat (water, propane).
     - Some liquids lose their lubricating qualities when heated (water).
     - Oils can varnish and carbonize with heat.
     - Some liquids can form a skin with heat (milk, paints).
     - Plated seal faces can suffer heat check with additional heat.
     - Acids become more corrosive when they are heated
2. Balanced seals can seal vacuum – this is common in condensate and lift pumps.
3. High Pressure – High pressure was the original purpose bringing about the development of balanced seals.
4. Less energy consumed – Saving energy is more important every day.
5. Less wear – All other conditions being equal, balanced seals wear less than their unbalanced versions.
6. High Speed Shafts – Shafts at 3,000 rpm or higher causes the seal faces to generate even more heat. Balanced seals reduce the pressure and force between the faces, thereby generating less heat.

7. Compensates Operational Practices – It's a common practice in many plants to close or throttle a discharge valve with the pump running to meter the flow through the pipes.
8. Pressure Spikes – They're inherent in the design of many systems.
9. Eliminate the Re-circulation line – A discharge bypass line is wasted energy and lost efficiency. Eliminate it with a balanced seal.
10. Less external flush – Less heat generated signifies less cooling requirements. Balanced seals can be flushed with as little as 1 or 2 gallons per hour.
11. No need to cool hot water – If the seal's elastomer can take the temperature, and the fluid is pressurized above its vapor pressure, the cooling line can be eliminated.

Balance explained by math

In the following illustration, the pressure inside the pump is 100 psi and the area of the seal exposed to the pressure is 2 in^2. Therefore, this seal is sealing a closing force of 200 pounds (Figure 13–9).

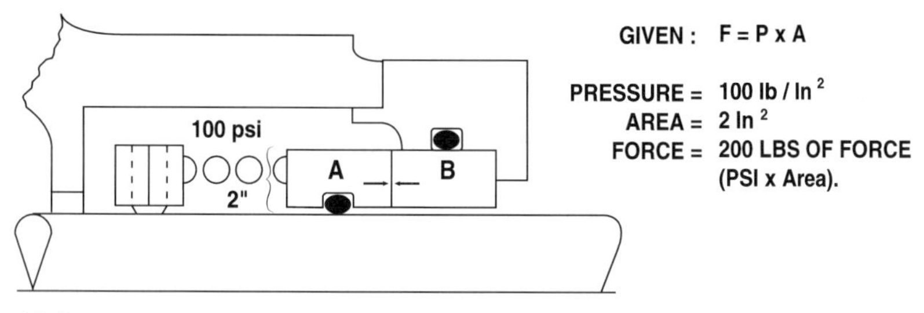

Figure 13–9

In the following illustration (Figure 13–10), we see that the pressure drops from 100 psig at the OD of the seal faces to 0 psig at the ID

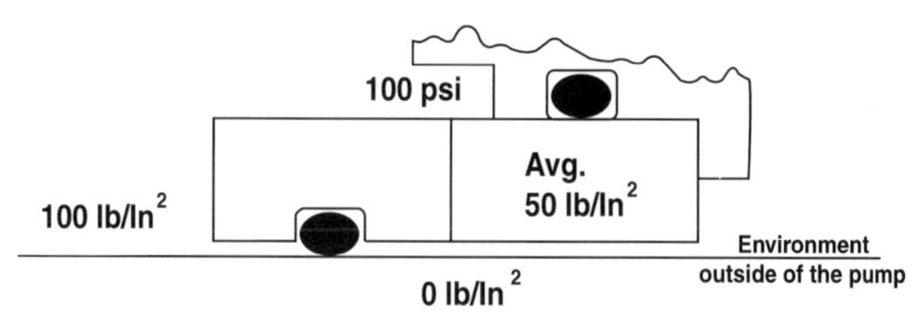

Figure 13–10

of the faces. Therefore the average pressure between the faces is 50 psig.

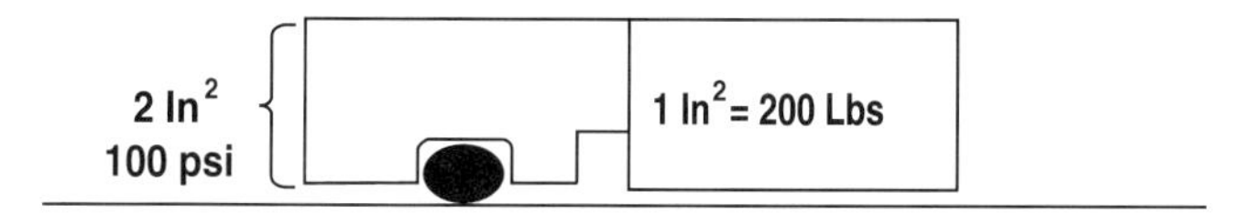

Figure 13–11

By varying the area of contact (in this case, projecting 2-in.^{2} of area exposed to pressure, over 1-in.^{2} of contact area) between the seal faces with a constant pressure, the closing force between the faces can be manipulated (Figure 13–11).

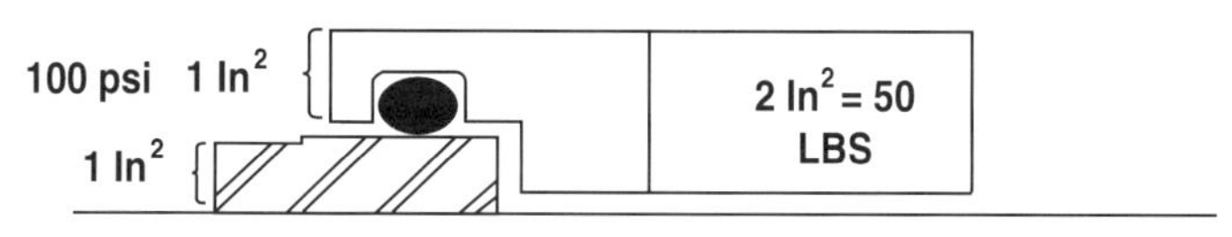

Figure 13–12

The illustration above (Figure 13–12) shows the same concept, but in the opposite direction. In this case 1 in.^{2} of area exposed to pressure is projected over 2 in.^{2} of contact area. The closing force is 50 lbs.

AUTHOR'S NOTE

A common example utilizing the same concept is a woman's pointed high heel shoes. She can leave imprint marks on a vinyl tile floor from impression of her shoe heel points. Another example would be the Eskimo, who can walk on soft snow without sinking by varying the area of contact of his footprint with broad snowshoes.

The pressure drop across the faces is not always linear. It may be convex or concave (Figure 13–13).

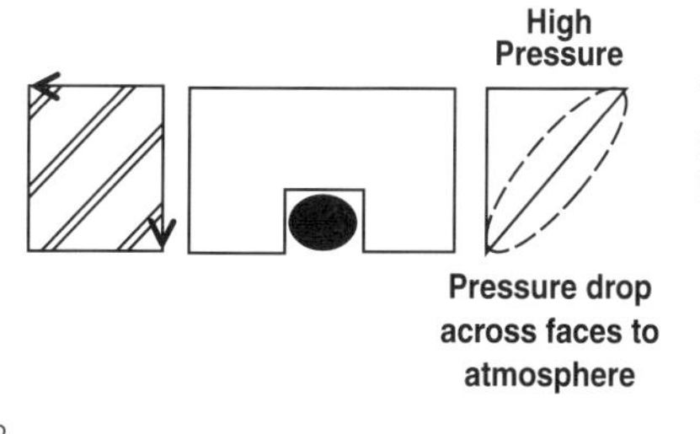

Figure 13–13

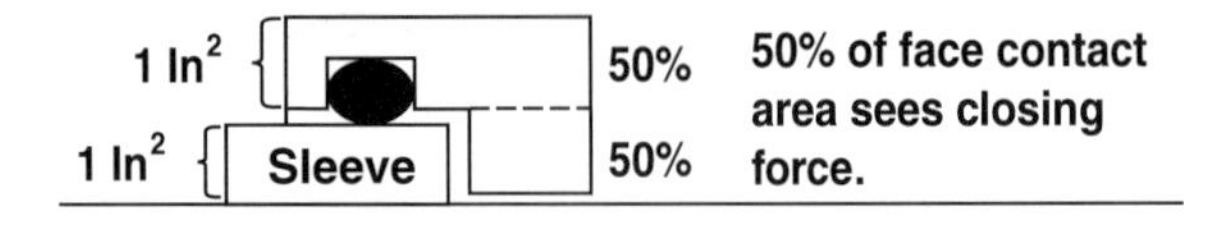

Figure 13–14

The illustration above (Figure 13–14) shows a perfect balance between opening and closing forces. This is not always convenient. A quick drop in atmospheric pressure, or loss of discharge resistance could cause the faces to open.

The following illustration (Figure 13–15) shows a more realistic balance ratio. 70% of the face sees closing forced and 30% of the face (falling below the stepped sleeve on the shaft) does not see closing force.

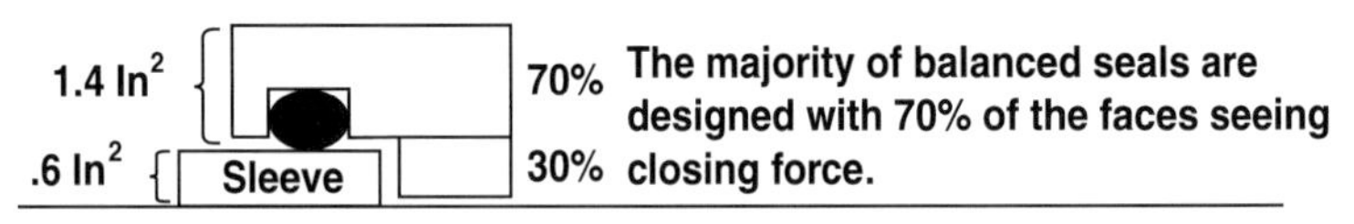

Figure 13–15

Here is a typical balanced rotary element of a mechanical seal with a 70/30-balance ratio (Figure 13–16).

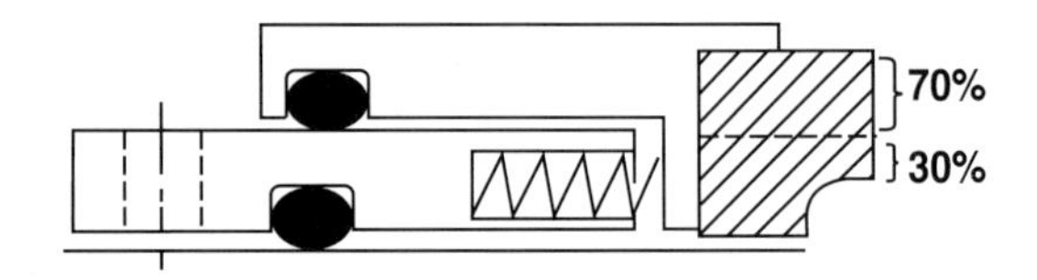

Figure 13–16

Figure 13–17 shows a balanced, single, rotary element seal mounted onto the pump shaft, pressed against the stationary face and gland, mounted in the seal chamber. Note the individual component parts of the rotary element.

Cartridge Mechanical Seals

Cartridge mechanical seals are designed so that the rotary and stationary elements, the springs and secondary seals, the gland, sleeve and all accompanying parts are in one integral unit. It installs in one piece instead of the numerous individual pieces.

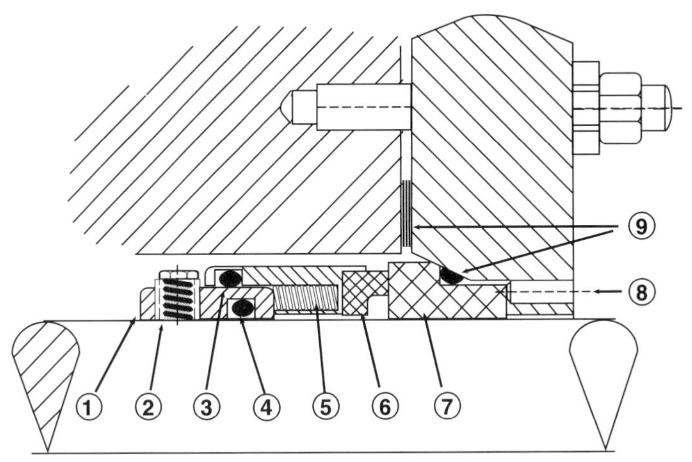

1. **Seal posterior (reference point for installation).**
2. **Set screw.**
3. **Secondary dynamic seal ('O' ring).**
4. **Secondary static seal ('O' ring).**
5. **Springs.**
6. **Rotary face.**
7. **Stationary face.**
8. **Anti rotation Pin.**
9. **Secondary seals.**

Figure 13–17

Advantages of the Cartridge Seal

- Pre-assembled at the factory.
- Reduced installation time.
- Can be installed on most pumps without total pump disassembly.
- Requires less technical expertise.
- The mechanics don't have to take complicated and confusing installation dimensions with respect to spring tension and face loading.
- The mechanics can't touch or contaminate the faces of the seal during installation.
- The installation costs less.
- The cost of the seal is less as a total unit than the sum of the individual parts.
- Some cartridge seal models have accompanying re-build kits.
- Most cartridge designs incorporate ports and hardware for connecting environmental controls such as:
  - Flush
  - Quench and Drain

– Disaster Bushings for API compliance

- Cartridge Seal designs comply with most pump standards like:
  - ANSI
  - API
  - ISO
  - DIN

Here is an illustration of a typical cartridge mechanical seal mounted on a shaft in the seal chamber of the pump (Figure 13–18).

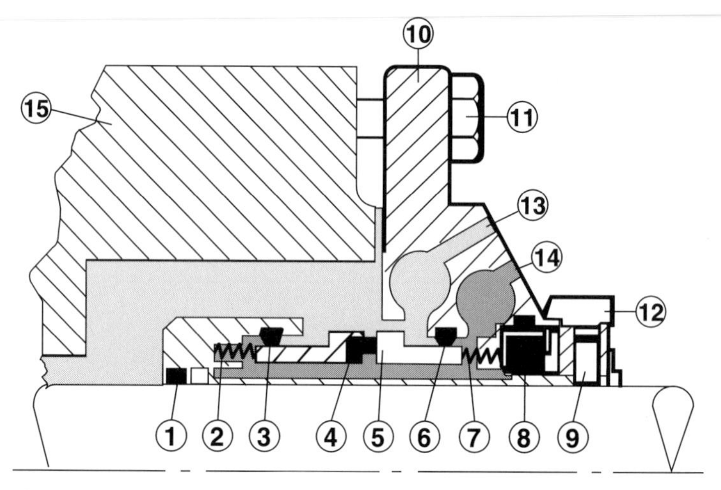

1. Static 'O' ring.
2. Spring.
3. Dynamic 'O' ring.
4. Rotating face.
5. Stationary face.
6. Static 'O' ring.
7. Self alioning springs.
8. Disaster bushing APi.
9. Set scren.
10. Gland.
11. Nuts.
12. Spacer for spring tension.
13. Flush line.
14. Cooling and draining.
15. Pump body.

Figure 13–18

Double seals

Double seals are also known as dual seals. They are used:

1. As an environmental control in difficult sealing applications.
2. To compensate for certain operational conditions like:
   - Operating a pump against a shut valve or 'dead heading' conditions.
   - Inadequate NPSHa.

- Air aspiration.
- Entrained gas bubbles.
- Turbulence.
- Dry-running the pump.
- Pumps in intermittent service.
- When operating just one pump in a parallel pump system.

3. On costly liquids.
4. With explosive liquids.
5. With toxic liquids.
6. With volatile liquids (tendency to gas or vaporize).
7. As the already installed spare seal.

Many plants use dual cartridge seals because:

1. They can be tested before accepting the seal into stock from the vendor.
2. They can be vacuum- and pressure-tested before installation.
3. In easy applications, these seals reduce maintenance costs by offering the life of two seals with only one installation.

The tandem dual seal

This dual seal has both the rotary units facing in the same direction. This type of seal is recommended for very high pressures. The support system, and thus the area between the two seals, would be pressurized at ½ the actual seal chamber pressure inside the pump (Figure 13–19, next page).

For example, if the application is actually 800 psi, you would pressurize the barrier tank support system and the area between the two seals at 400 psi. This way the inboard seal would seal 400 psi (800 – 400 = 400 psi) and the outboard seal would also seal 400 psi (400 – 0 = 400 psi). Each seal independently could seal maybe 500 psi, but not 800 psi. Together, the two seals in this tandem arrangement can seal 800 psi and higher depending on the barrier tank pressure.

Some purists hold back from calling this type seal a true 'double or dual seal'. This is because if one seal in this tandem arrangement fails, the other will immediately fail too.

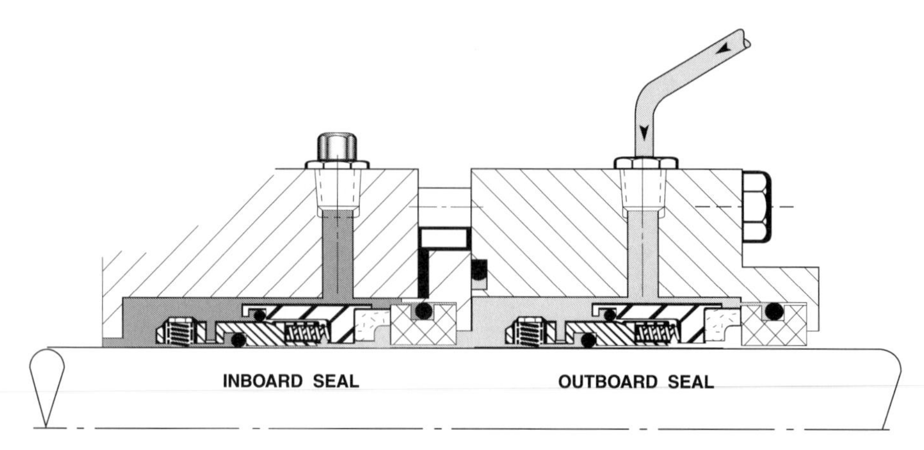

Figure 13-19

AUTHOR'S NOTE

The general tendency with double seals is to think that one seal in a dual arrangement could fail and the other would kick-in as a spare or back up. This is why cars have a spare tire, why a passenger jet has a pilot and a co-pilot and why our bodies have two kidneys and two eyes. One could become incapacitated, damaged or fail and the other would assume the functions. This is the case with other double seals but not this one in the applications where it is most specified. As soon as one fails the other will follow.

The tandem dual seal is mostly mated to a pressurized barrier tank, or a pumping unit as the support system. Support systems are discussed later in this chapter.

The back-to-back double seal

This type of double seal (with back to back faces) is pressurized above the pressure inside the seal chamber (Figure 13–20). It is recommended for toxic, explosive, costly, dangerous, and volatile liquids. It is important to maintain the seal pressure above the pumped pressure inside the seal chamber.

To obtain the benefits of this seal, it is necessary to install a gauge indicating the actual seal chamber pressure. Sensors and transmitters can be used to monitor and act on a pressure change. One of the two seals can fail without product loss or fugitive emissions. This seal would be connected to a pumping unit seen later in this chapter.

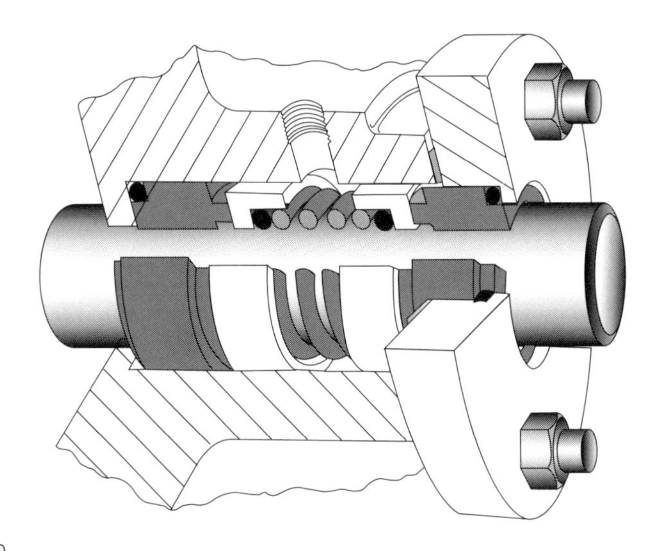

Figure 13–20

The face-to-face dual seal

This face-to-face double seal is quite versatile (Figure 13–21). It is recommended in a wide range of applications depending on the piping arrangement of its support system.

With its support system pressurized above the seal chamber pressure, this double seal functions well with toxic and dangerous liquids (like the back-to-back dual seal). If the support system is pressurized at ½ the seal chamber pressure, this seal can handle higher pressures where a single seal would fail. If the support system should be non pressurized

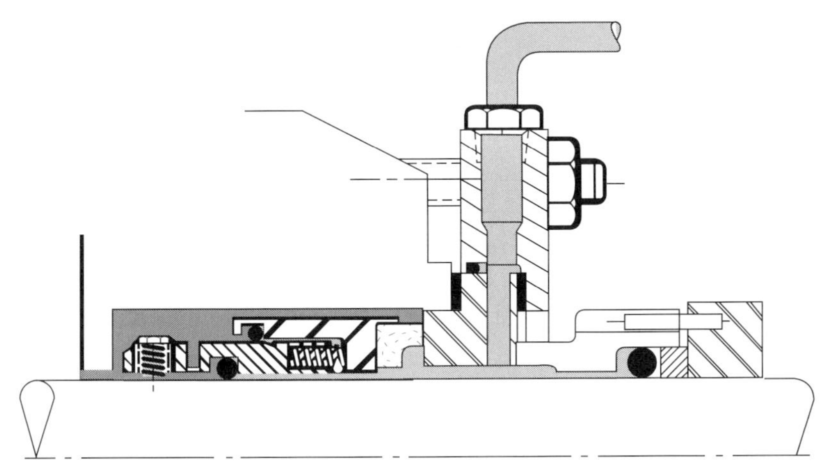

Figure 13–21

with forced flow (not induced flow), it becomes a good seal for resolving some operational and design problems in the pumping system like air aspiration, inadequate NPSHa, and operation away from the pump's BEP. Let's look at support systems.

Support systems for dual seals

Double seals require some type of support system. The reason is simple. With two seals mounted onto the same shaft, one seal is the principal or primary seal and the other becomes the secondary or back up seal already installed. If the primary seal is performing its function and sealing the pumped liquid, the secondary seal would be running dry, overheat, burn and self-destruct. Then when the crucial moment comes, we won't have a second seal to assume the functions, which was the original reason to consider a dual seal.

So the support system serves to lubricate and cool the faces of the secondary seal while the first is performing its functions. The pressure applied to the support system governs the optimum life of the dual arrangement, and it also governs what happens when one of the two seals fails.

There are three distinct support systems for double seals. They are often referred to as 'barrier tanks'. The term barrier tank was initially applied to the thermal convection tank, although the term today refers to any of the three support systems. Each support system has different attributes.

The thermal convection tank

The Thermal Convection Tank (Figure 13–22).

- Can be pressurized or un-pressurized.
- Conducts a re-circulated flow by thermal convection. When enough heat is generated inside the mechanical seal, it expands initiating a flow into the tank where it cools and contracts and is brought again into the seal.
- The tank is sealed and welded, and meets the boiler code for pressure vessels.
- Has a specific location in relation to the seal to optimize the convective flow.

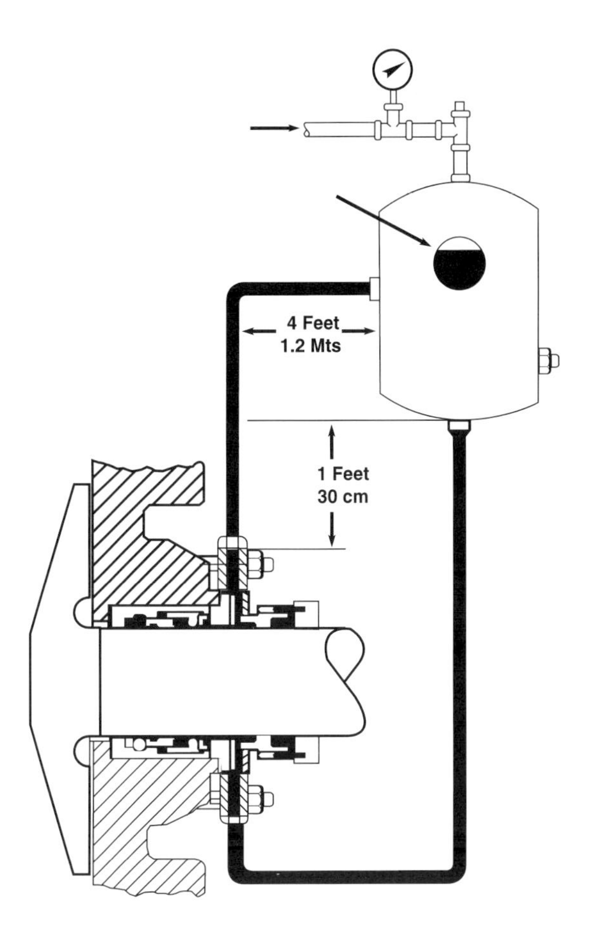

Figure 13–22

The turbo tank

1. Is a low pressure system.
2. Generates its own forced flow with a submersible centrifugal pump.
3. Is ideal for surviving operational problems like:
   - Cavitation
   - Inadequate NPSHa
   - Dry-Running the pump
   - Dead-heading the pump
   - Operation away from the BEP on the curve
   - Vacuum

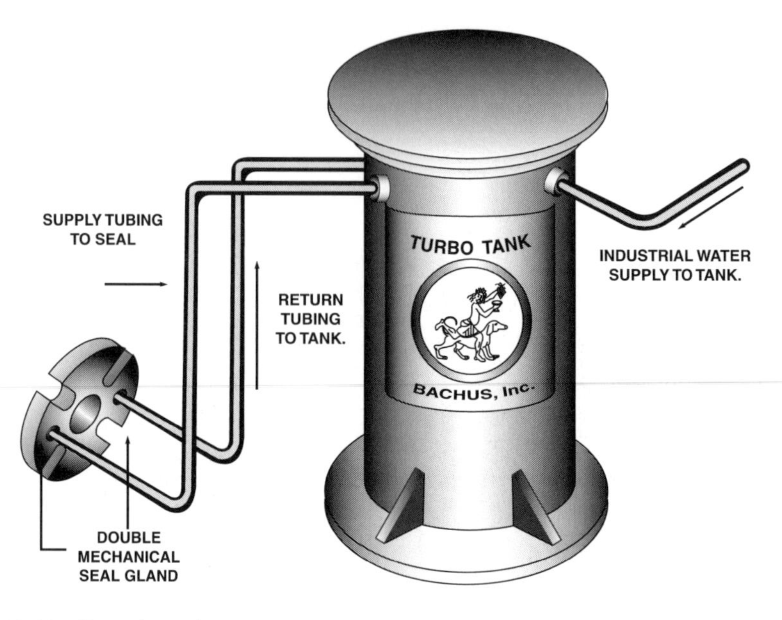

Figure 13-23 The turbo tank

- Intermittent service
- Air aspiration
- Turbulence
- Parallel pumping.

4. Is good as an environmental control (to reduce heat, dissolve crystals, absorb gases).
5. Is ideal as the installed spare.
6. Continues cooling independent of the process pump.
7. Is not location specific in relation to the seal.
8. Is not a sealed pressure vessel.

The pumping unit

- Works with water or oil as a barrier fluid.
- Is not a sealed pressure vessel.
- Generates its own re-circulation flow and pressure with a PD pump.

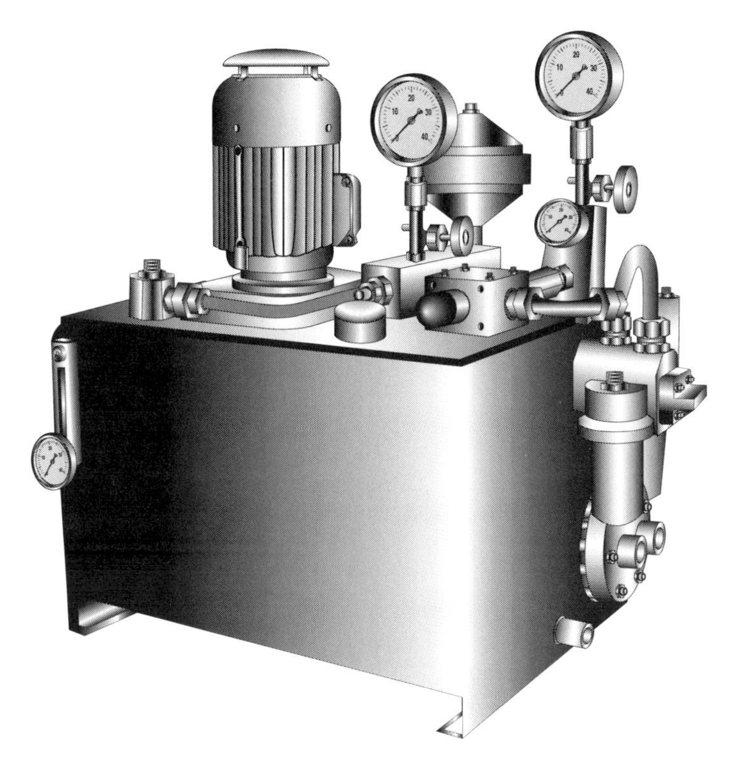

Figure 13-24 The pumping unit

- Incorporates flow and pressure regulators, an in-line filter, and an internal heat exchanger.

Failure Analysis of Mechanical Seals

Causes of premature seal failure

There are numerous reasons why seals fail prematurely. The origin of the failure can reside in the pumping system, in the pump operation, or the maintenance shop, the storeroom, or even before the seal arrived into the plant. The first sign of failure normally is liquid on the floor.

It's possible that a failure occurred to the seal at the manufacturer. Nowadays, the companies that we call 'Manufacturers' or 'Factories' are actually 'Parts Assemblers'. It's likely that no mechanical seal manufacturer is really fabricating their springs, seal faces, or set screws. These 'Assemblers' contract other companies to fabricate the pieces. The failure could have originated in the moment that a sub-contractor made a piece. The failure could have originated on the seal assembly line. The majority of the 'Seal Assemblers' would perform a static pressure and vacuum test on their final product. It's unlikely that they would perform a dynamic pressure and vacuum test on their creations. So if an anti-rotation pin were left out at the factory, it's likely this would not be identified until the seal and pump are started. And although these 'Assemblers' don't want to admit it, there are actually deficient seal designs being delivered to the customers that simply don't work.

The storeroom techs must protect the mechanical seals handling them with extreme care. The engineer or technician must specify and select the seal components correctly. The mechanical maintenance technician must protect the seal at the moment of installation. Once the seal is installed into the pump, the seal requires the correct environmental controls to assure optimum life. At this point the work of the maintenance technician ends.

We could say that if the mechanical seal fails immediately, or within moments of the pump start-up, one should investigate the events before the start-up. This failure is probably in the installation, or handling or manufacture of the seal. If the seal fails in a few days, the failure might be an incorrect specification of a component like an o-ring seal. But if the mechanical seal fails after three weeks, or 2 or 7 months of service, now we must consider the operation and/or design of the system.

On starting the pump and motor, the operators control the service of the mechanical seal. The operators and the process engineers have a tremendous influence on the optimal life of the mechanical seal, just as the operator of a car has the most influence over the optimal life of his automobile. The pump must be operated at, or close to it's best efficiency point (BEP) on the pump curve.

If the pump is operated away (to the left or right) from it's BEP on the curve, the pump will vibrate. This damages the bearings and the seal faces leading to premature failure. Also, operation to the left of the BEP on the pump curve adds more heat to the fluid, which can damage the o-rings within the seal. In severe cases the fluid can vaporize leaving the seal to run dry without cooling or lubrication. This damages the seal. And if the pump is operating to the right of the BEP on its curve, besides the vibrations, the poor pump can go into cavitation, and this certainly will kill the seal. If a person insists on mistreating his car, driving his car like bumper cars at the circus or fair, it is not the fault of the auto mechanic if he can't maintain the car in good operating condition.

AUTHOR'S NOTE

It has always seemed strange to us that the mechanics, or the manufacturer, are blamed when a mechanical seal fails after 3 months of service. If the seal fails on start-up, maybe you could point to the mechanic or the seal, but not after 3 weeks or 4 months of operation. This would most likely be an operational failure (a failure in operations), or a design failure (a failure in the system's design). And what is really amazing is that this statement and these words have never been recognized or said before.)

There's a need to introduce some logic to mechanical seal failure, and all pump failures. When an operator sees a pump leaking and dripping through the seal, he blames the mechanic, and the mechanic blames the seal manufacturer. The seal manufacturer blames the pump manufacturer. The pump manufacturer blames the plant's purchasing agent. The purchasing agent blames the engineer, and the engineer blames the operator. Now we've gone all around the block just to get next door. I suppose we could say that when an orchestra gives a bad concert, then no one is to blame and everyone is to blame.

When a seal is installed into a pump, and the motor started, an imaginary line is drawn, and the seal begins a journey toward the day when the seal will come out of service, either from premature failure, or from obtaining its maximum service life. On one side of the imaginary line are the events in the manufacture of the seal, its handling, storage, and installation. On the other side of the imaginary line are the events that occurred after the pump and seal were started the first time.

AUTHOR'S NOTE

Maybe you've heard your auto mechanic say on installing a new radiator water pump or alternator onto your car ... that if it runs for 5 minutes, then it will run for 15 years. If it's going to fail, it will do so within the first five minutes.

Mechanical seal problems originating in the factory, storage, handling, and installation will be evident within the first few moments or hours of operation. Consider fractured faces (from poor handling), or a missing o-ring (from poor assembly), or installing a 50 mm seal onto a 48 mm shaft (poor installation).

These failures and the leaks will be immediately evident.

If there is a chemical incompatibility between the liquid and an o-ring rubber compound, or if heat is generated from too much spring tension, this will be evident within a few hours or days. Galvanic corrosion or inadequate spring tension will reveal itself in a few weeks.

Certainly, at the moment of starting a pump with a new seal installed, the events prior to the installation begin to disappear as a cause or origin of failure, and the factors of operation, process, and design in the system begin to appear as possible reasons for any premature failure.

Let's begin the analysis, or autopsy with the physical evidence on the component seal parts.

O-ring (The elastomer) failure

About half of all pumps in the shop today were pulled out of service because they were leaking or wouldn't hold pressure or pump. This is most likely a leaking o-ring. The o-ring is the rubber component of most mechanical seals. The o-ring controls the temperature, pressure, and chemical resistance of the mechanical seal (Figure 14–1).

The difference between a mechanical seal in a pump in alcohol service and a pump in steam service is the o-ring. It is not the stainless steel, or the ceramic face of the seal. The difference between a mechanical seal in

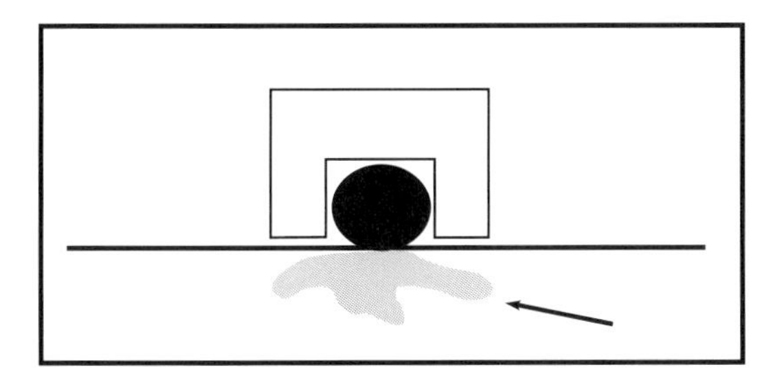

Figure 14–1

ammonia service, and a mechanical seal in propane service, is the o-ring. The people who assemble seals install o-rings that are adequate to perform the static pressure and vacuum test, which is normally done with water or air. The user must verify that the seal elastomers, the o-rings, installed in the factory are adequate for the service application (temperature, pressure, and chemically compatible). If they are inadequate, they must be exchanged for the correct o-ring rubber compound before the installation. It may be necessary to use a tool, like the following, to identify the o-ring elastomeric compound (Figure 14–2).

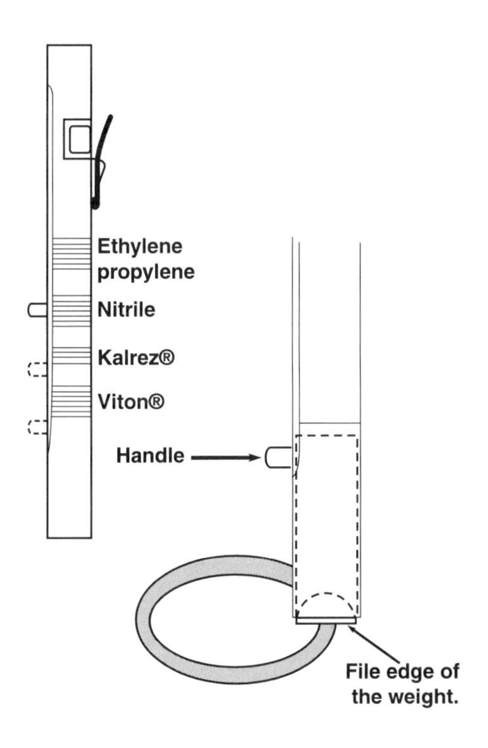

Figure 14–2

Common o-ring compounds used in mechanical seals

There are many rubber compounds used in industry as o-ring secondary seals. Some elastomeric compounds are only found in one or two sealing applications in one industry. We have listed the following four compounds, which find broad popularity in almost any production plant.

1. **Fluorocarbon (Viton®)** is a rubber compound that is compatible (meaning it resists without degradation) with most petroleum based liquids and gases (propane, gasoline, crude oil), some acids and other chemicals. It is used extensively in the petroleum refining and petrochemical industry. Its temperature range is good from –15° F to +400° F. (–25° C to +205° C).
2. **Perfluorocarbon (Kalrez®, Parofluor®, Chemraz®)** is a rubber compound compatible with most organic and inorganic liquids and gases and aggressive chemicals. This material finds popularity in chemical processing and pharmaceutical plants, and wherever the temperature of the application demands. Its service range is from about –20° F to +500° F. (–30° C to +260° C).
3. **Nitrile (Buna-N)** is a rubber compound popular in most household plumbing applications. It's a basic plumbers o-ring seal, and handles most household liquids and chemicals. Because industry pumps so much water, this elastomer may be the single most popular o-ring secondary seal in the world. Its service range is from –30° F to +250° F (–34° C. to +120 C).
4. **Ethylene Propylene (EP, EPDM)** is an o-ring rubber compound that is compatible with most water-based chemicals. It is good with caustic soda, detergents, water treatment chemicals, steam, and wastewater and with food processes like milk, beer, and soups. EP rubber compound is petroleum based and for this reason it should never come into contact with petroleum based chemicals.

   It will dissolve. Its service range is from –70° F to +300° F (–57° C to +150° C).

How many different o-rings to heat some water?

Let's consider an industrial boiler. You may need at least three of these previous mentioned o-rings just to prevent leaks and drips in a simple hydronic or steam boiler. Raw water comes into the boiler room with pipes, gauges, valves and instrumentation. All these fittings would probably use Nitrile rubber o-ring seals to give long-term leak free service.

Next, the raw water must be treated before it can be pumped into the boiler. Treating the boiler water does three things. First it controls the pH so that the boiler tubes won't corrode. Next the treatment process removes oxygen, which prevents internal boiler wetted parts from rusting. Third, the treatment process removes minerals from the water so that mineral scale won't form on the boiler tubes, insulating them, and causing the boiler to lose efficiency. After raw water has been treated with chemicals to scavenge the oxygen, remove or neutralize minerals, and control the pH, the raw water becomes make-up water. The treatment chemicals and the treated water will need Ethylene Propylene o-rings on the mechanical seals, instrumentation, valves, connections and fittings.

If the boiler is a high-pressure boiler, the boiler's discharge valves, and instrumentation fittings may need perfluorocarbon o-rings for temperatures above 300 degrees. The high-pressure boiler feed water pump may need these high temperature o-rings in the mechanical seals because of the high frictional heat generated by the seal faces. If the DA (deaerator) tank is sealed and pressurized to hold the hot water from flashing, it may need these high temperature o-rings.

If the boiler burns propane, natural gas, or fuel oil, then you'll need fluorocarbon o-rings on your fuel lines, valves, instrumentation and fittings.

Who would have thought that an industrial boiler would need up to four different o-ring compounds just to heat some water?

When an o-ring comes out of service, it should be inspected for signs of damage and degradation. These could be:

Chemical attack

Because the o-ring comes into contact with the fluid, the o-ring's rubber compound must be chemically compatible with the fluid. Chemically compatible means that the o-ring will resist the chemical without degradation. If the o-ring is not chemically compatible with the pumped liquid, it may swell, harden, dry and crack, soften, or even dissolve depending on the nature of the chemical attack. The surface of the o-ring may form blisters, scale, or form fissures and cracks. The cause of these symptoms generally is chemical attack. The attack may come from the pumped liquid, or from the barrier fluid or external flush in the case of installed environmental controls. Environmental controls are discussed later in this chapter. Too much heat in the system may present the same evidence. You must be familiar with the different o-ring compounds used as secondary seals in mechanical seals, and all instrumentation, connections and fittings. You must know the temperature limits (upper and lower) and the chemicals they are compatible with.

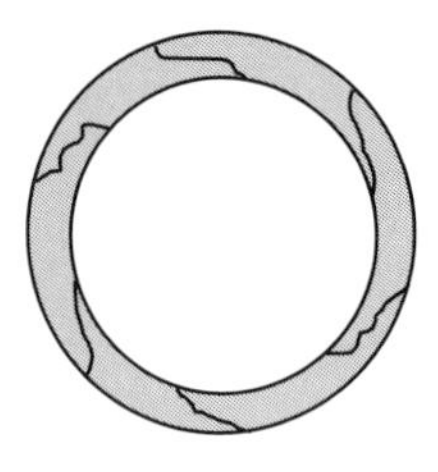

Figure 14-3 Chemical attack

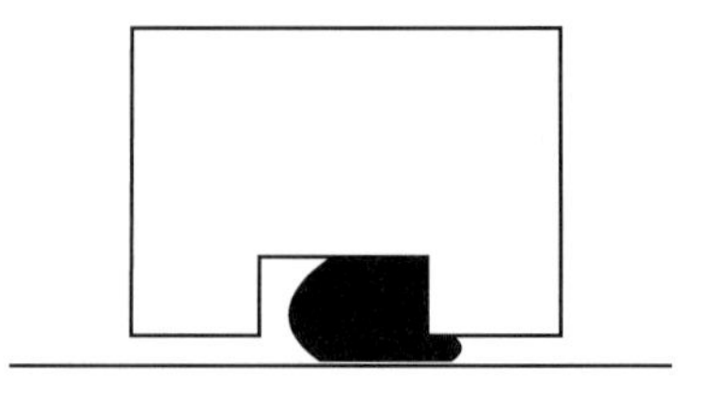

Figure 14-4 Extrusion

Extrusion

Extrusion is deformation under pressure. An o-ring extrudes when the pressure is too high. Maybe the o-ring needs back-up rings to tighten tolerances. Maybe the design of the o-ring groove is inadequate. Some o-ring compounds get softer as the temperature rises so temperature-linked-to-pressure is also a factor to consider. An o-ring also may soften if under chemical attack. This should also be checked as a possible source of the extrusion.

Split, cracked or hard o-ring

Although an o-ring softens with temperature, too much heat will harden the o-ring. Too much heat is the usual cause, but chemical attack may produce the same result. The o-ring must deal with the temperature of the application and also from the heat generated between the mechanical seal faces. Many mechanical seal designs place the o-ring where it receives a lot of heat. A good idea is to find a mechanical seal with the o-rings and other secondary elastomer seals placed away from the heat of the faces.

Compression set

This is a good indicator that the O-ring was exposed to too much heat. Compression setting means that the round cross-sectioned o-ring came out of its groove with a squared cross section. The heat caused the o-ring to re-cure in the groove, taking the shape of the groove.

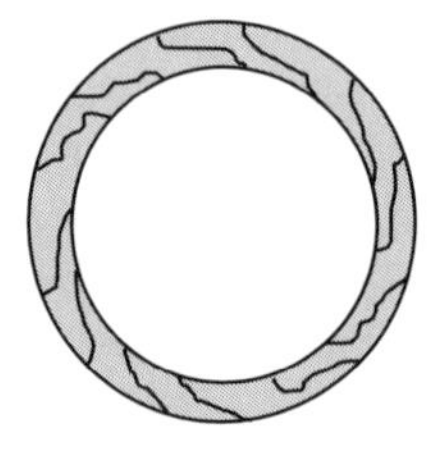

Figure 14-5 Split, cracked or hard o-ring

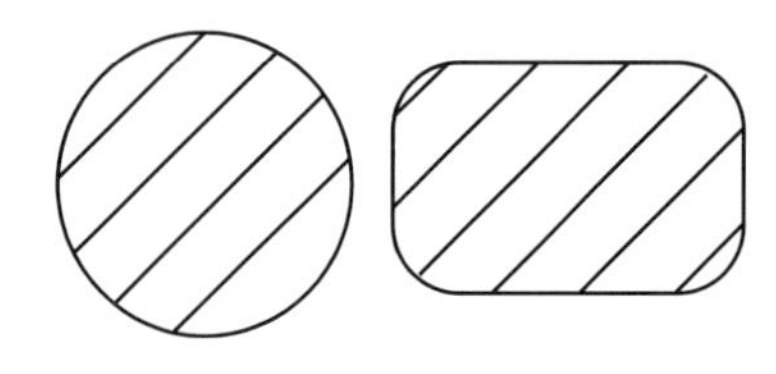

Figure 14-6 Compression set

Ozone attack

O-rings, especially Buna-N (Nitrile compound), should be stored away from fluorescent lighting and electric motors. These are sources of ozone. Ozone causes a general degradation of these elastomers.

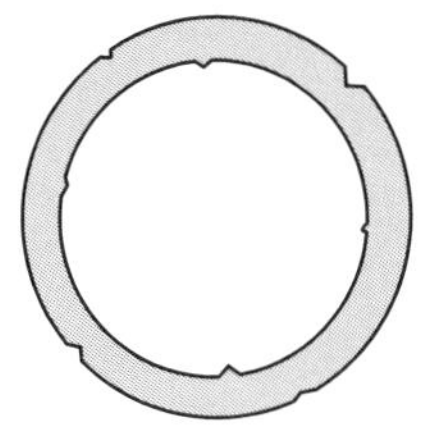

Figure 14–7 Nicks and cuts

Nicks and cuts

This failure is mostly evident on start-up as immediate leakage. Extreme care should be taken to prevent damaging the o-rings as the mechanical seal is slid down the shaft at installation. The o-ring grips the shaft to withstand the seal's maximum pressure rating. The o-ring must slide over the impeller threads, key way grooves, steps on the shaft, and marks made by previous setscrews. O-rings damage easily.

AUTHOR'S NOTE

Most pumps don't come out of service because they break. Most pumps go into the shop because they're leaking. Every day all over the world, too many $2,000.00 mechanical seals are thrown into the recycle bin or thrown away because of an 11¢ o-ring.

The elastomer sticks to the shaft

The elastomer or o-ring is seated inside the mechanical seal. It must be free to move and flex with the seal (Figure 14–8).

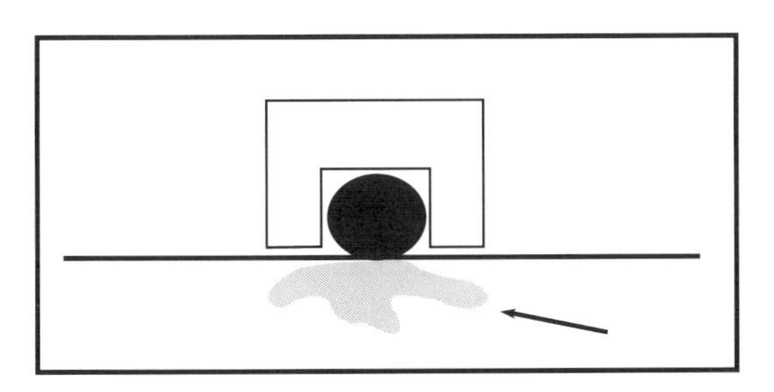

Figure 14–8

The pump shaft must be free to slide through the o-ring. However, crystallizing liquids, accumulated solids and pastes, and products that harden can cause the o-ring to hang-up or stick to the shaft losing its freedom. Oversized shafts and sleeves can also be blamed. If the seal is misaligned onto the shaft the o-ring can fret the shaft and hang into its groove. Heat and chemical attack can also make the o-ring vulcanize to the shaft. Any of these can occur. As the shaft moves within the tolerance of the bearings, it can drag the seal faces open or crush them together if the o-ring sticks to the shaft.

The springs clog and jam

Suspended solids, crystals and sediment in the pumped liquid can lodge into the seal springs and restrict their movement. Jammed springs cannot flex to maintain the seal faces united while the shaft moves within the bearing axial tolerance. It's best to use seals designed for slurries and solid particles. Many of these designs have the springs placed outside the pumped fluid. The seal chamber design and piping also has an influence on the seal's ability to survive while handling solid particles. Seal chambers designed with tapered, spiraled, and open bores facilitate the handling of suspended solids (Figure 14–9).

The discharge re-circulation line is a dinosaur of design, held over from the days when pumps had packing. With packing, the discharge bypass line prevented the entrance of air through the packings on pump start-up, and also provided some cooling to the packing rings. With a mechanical seal in slurry service, this discharge bypass line blasts the mechanical seal and chamber with the highest concentration of solids in the pump. It will destroy the seal in short order. If some cooling and flushing is desired, a suction bypass line, from the seal chamber to the pump suction nozzle, is preferred over a discharge bypass line, from the pump discharge to the seal chamber.

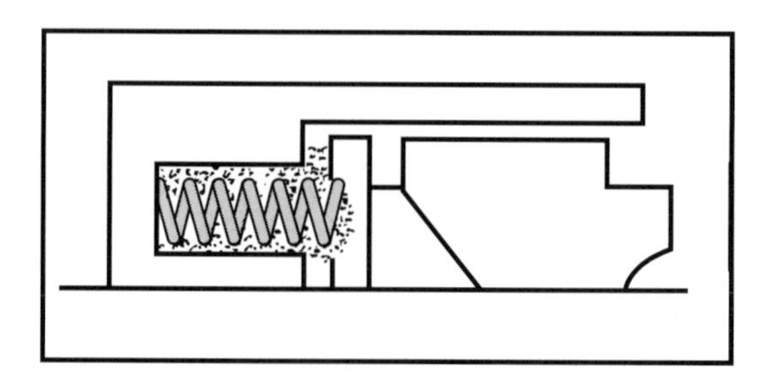

Figure 14–9

Figure 14–10

The shaft frets under the shaft seal

If the mechanical seal is cocked or misaligned onto the shaft (Figure 14–10), or the seal chamber face is not perpendicular to the shaft, the shaft o-ring will have to flex to maintain face contact. The flexing causes the shaft seal to rub on the shaft under the seal, which can eat or erode a groove into the shaft. The pumped fluid can leak under the o-ring, or the o-ring can hang into the groove and drag the seal faces as the shaft moves within the bearing tolerance. This is considered a seal problem, but is actually an alignment problem. There is more information on this later in this chapter regarding pump reconstruction. Figure 14–11 shows how an o-ring in a misaligned seal can fret the pump shaft.

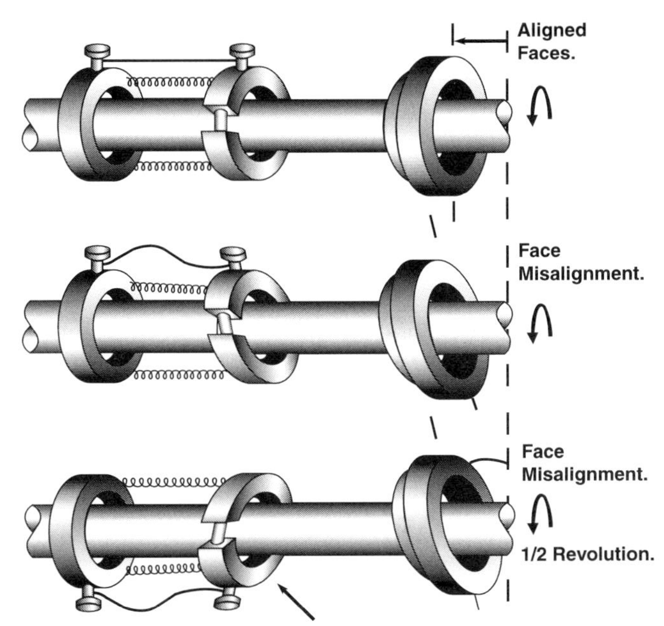

Figure 14–11

The o-ring seal gripping the shaft moves

- 1,800 rpm × 2 movements/revolution = 3,600 movements/minute.
- 3,600 movements/min × 60 minutes/hour = 216,000 movements/hour
- 216,000 movements/hour × 24 hours/day = **5,184,000 movements per day**

At 5,184,000 movements (rubs) per day on a shaft spinning at 1,800 rpm the constant friction will eat a fret mark (groove) into the shaft or sleeve in just a few days (Figure 14–12). The next o-ring seal, installed onto this pump, will ride in the groove (cut by the previous seal) and never give good service. Again, this is an alignment problem and not a seal problem. Many seal companies have addressed fretting corrosion with a product called self-aligning faces. Both the rotary and stationary faces are spring loaded as the faces push against each other. The opposing springs tend to cancel themselves and the union between the seal faces will always be perpendicular to the shaft axis. This prevents the flexibly mounted shaft seal from dancing and rubbing on the shaft if the pump parts should be out of alignment. Another way to resolve fretting corrosion is to align the pump parts upon rebuilding the pump.

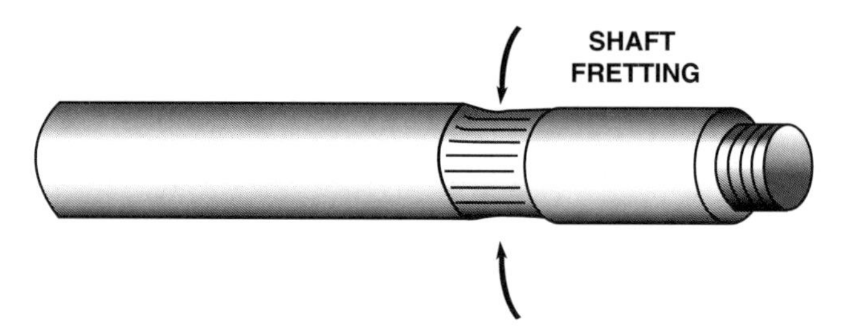

Figure 14–12

Incorrect installation dimension

Seal faces loaded with over-compressed spring tension, can generate too much heat. This will damage the o-ring and even fracture the seal faces from thermal expansion (Figure 14–13). If the spring tension is inadequate, the faces can leak after a short while as the softer face wears against the harder face, and the spring tension relaxes completely. The installation dimension is critical to long seal life.

Most seal companies provide a set of instructions and engineering schematic with each mechanical seal to help the mechanic determine the correct installation dimension. These instructions assume:

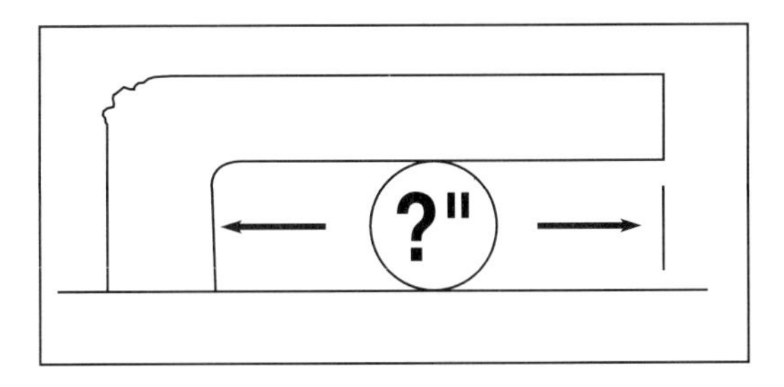

Figure 14-13

- That the mechanic has the drawings available at the moment of installation. Often times, the engineer stores the drawings in a file so they won't get lost or dirty. For innumerable reasons, the mechanic rarely has a set of drawings available at the installation.
- That the mechanic knows how to interpret engineering schematics.
- That the instructions and schematic are printed in the mechanic's native language.

Also the shaft or sleeve should measure the correct dimension within tolerance, and have the correct surface finish:

- The correct dimension tolerance =: +0.000:–0.002 inch. For example, a 2 inch diameter shaft can measure no greater than 2.000, and not less than 1.998 inches.
- The surface finish = 32rms, also known by machinists as ΔΔΔ

Environmental controls for difficult sealing applications

The easiest way to prevent many unplanned premature seal failures is to use balanced o-ring cartridge designed seals. Most of the seal manufacturers make models incorporating these design features.

Rarely is a balanced o-ring cartridge seal supplied as standard equipment with a new pump. It's thought that they cost too much and will raise the sale price of the pump.

Although the balance feature, o-rings, and the cartridge concept were discussed in detail in the previous chapter, here is a brief review why this seal design will give your pumps their best chance for extended leak-free service with reduced maintenance costs.

- Balance – indicates less heat generated. Heat is the principal enemy of all mechanical seals. There simply is no logical reason for specifying unbalanced seals.

- The O-ring – is the most widely available and inexpensive elastomer seal in the world. Practically all rubber compounds are available in an o-ring configuration. You can buy them from specialty houses or from the corner hardware store.
- The Cartridge Concept – has every component part of the sealing system in one unit. They're easier and faster to install and don't require total pump teardown. Most come with additional control ports and rebuild kits.

For most applications, the balanced, o-ring cartridge seal will adequately handle every pump, liquid, and condition in a modern industrial process plant. There are, however, some industrial pumping applications that will present problems to even the best of mechanical seals. Should one of these applications cause the seal to give less than desirable performance, the next step to take in extending the service life of the seal (and ultimately the pump) is to install some type of environmental control to protect and isolate the seal components from the fluid. Let's consider some difficult sealing applications.

Difficult pumping applications for mechanical seals

- Crystallizing Liquids
- Solidifying Liquids
- Vaporizing Liquids
- Film Forming Liquids
- High Temperature liquids
- Non-lubricating Liquids
- Dry-Running the Pump
- Dangerous Liquids
- Gases and Liquids that Phase to Gas
- Slurries (Suspended Solids)
- Cryogenic Liquids
- High Pressure Liquids
- Vacuum Conditions
- High Speed Pumps

The following pictures show different arrangements of Environmental Controls. These Controls expand the operational range of the mechanical seal, improving the pump service life.

Environmental controls

Suction bypass

The suction bypass is good for:

- Removing heat
- Removing suspended solids

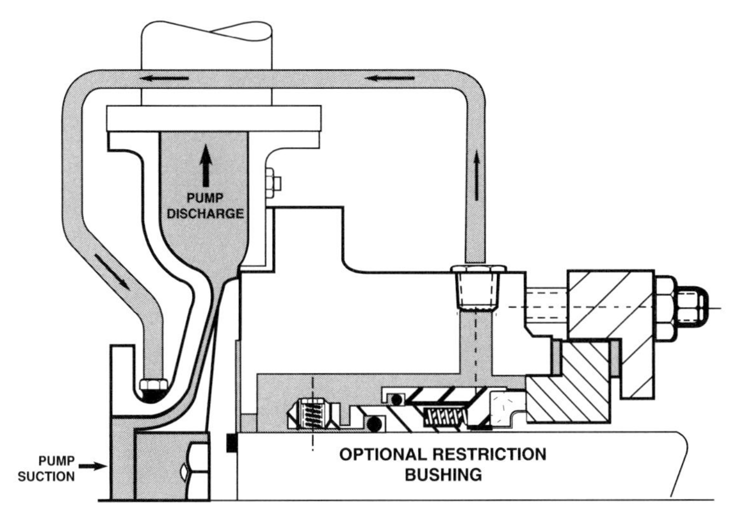

Figure 14–14

AUTHOR'S NOTE

Be aware that the liquid may vaporize if the fluid's vapor pressure should surpass the absolute pressure in the seal chamber.

The external flush

The external flush is good for:

- Removing heat
- Separating the seal environment from the pumped liquid

AUTHOR'S NOTE

This environmental control may also dilute the pumped product. If the external flush is water, it may be necessary to evaporate it later in a costly posterior process.

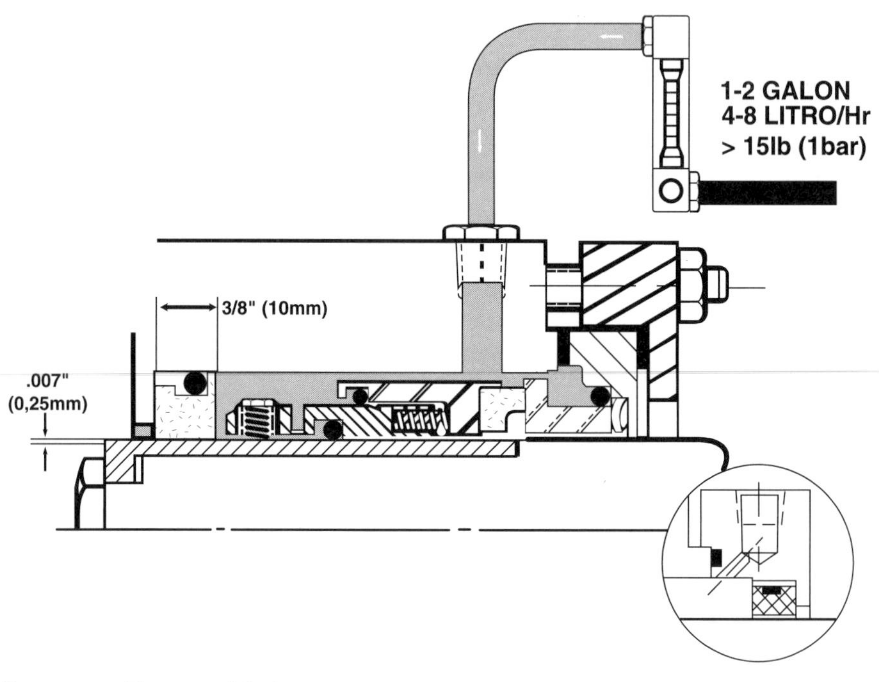

Figure 14-15 The external flush

The thermal jacketed seal chamber

The thermal jacket seal chamber is good for removing/controlling heat without introducing additional liquid into the process.

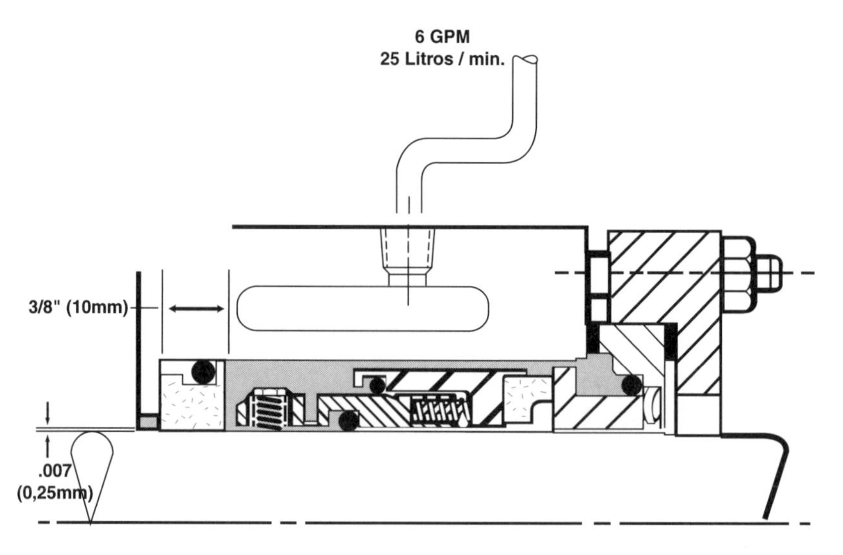

Figure 14-16 thermal jacketed seal chamber

The slurry seal with evacuation line

The slurry seal with the evacuation line is designed to handle/evacuate suspended solids, crystals, sediment, and dirt in the pumpage. The seal's springs are located out of the fluid. The o-rings move and rub across a clean surface as the faces wear. The o-rings are placed away from the heat generated by the faces.

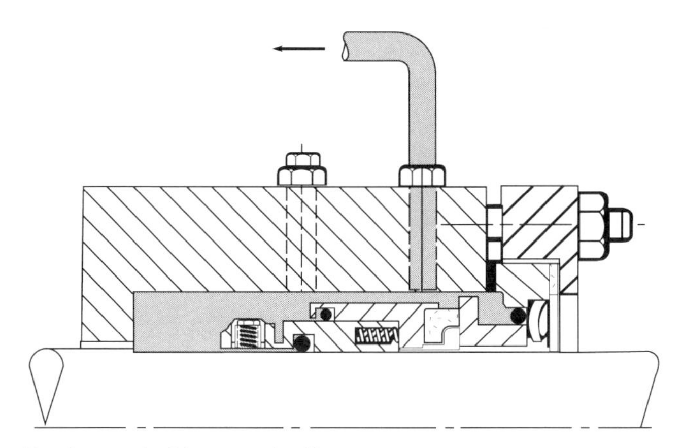

Figure 14–17 The slurry seal with evacuation line

The quench and drain

The Quench and Drain control is good for removing heat without contaminating or diluting the pumped product.

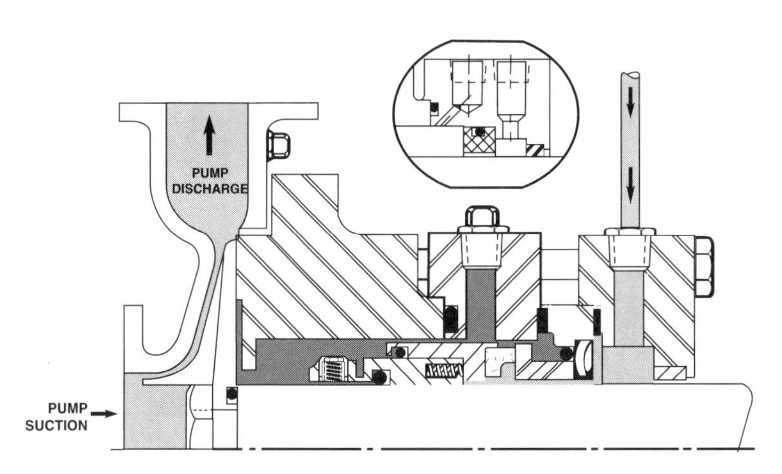

Figure 14–18 The quench and drain

AUTHOR'S NOTE

This method requires a source of water or steam for the quench. The pumped liquid may contaminate the drainage. If so, it should be discarded.

The heat exchanger

The heat exchanger is good for controlling temperature in the seal chamber utilizing the same pumped product. This method doesn't dilute the product.

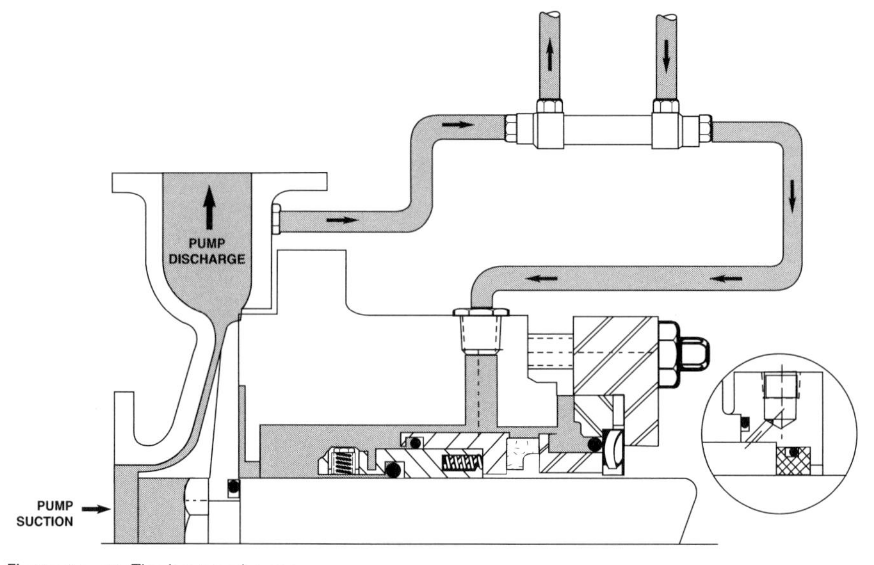

Figure 14–19 The heat exchanger

The double or dual mechanical seal

The dual seal is good to control the environment inside the seal without diluting the pumped product. It works with re-circulated barrier fluid. It establishes a barrier between the pumped fluid and the environment (Figure 14–20).

You've just seen six different concepts to control the environment inside the seal chamber and mechanical seal. Some methods are economical. Others are costly. Some have secondary side effects to contend with. Now let's consider the 14 difficult sealing situations and apply the environmental controls to extend the running time of the seal and pump.

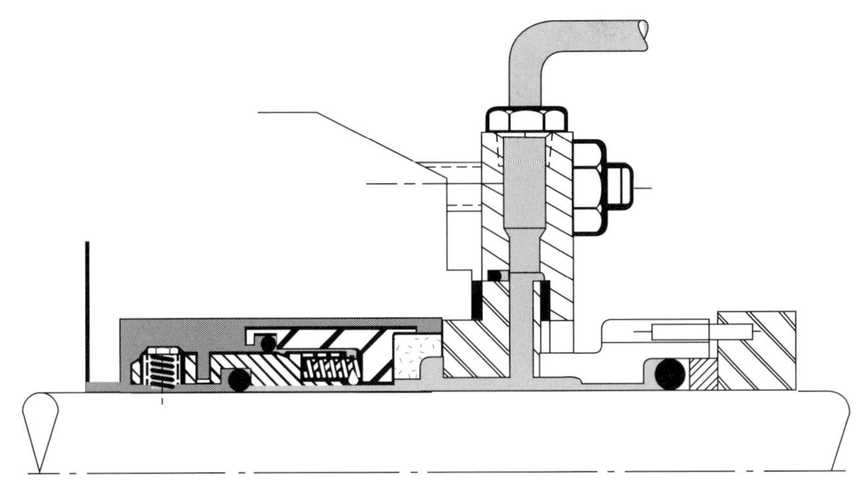

Figure 14–20 The double or dual mechanical seal

1. **Crystallization** Some liquids form crystals with heat (think of sugar and rock candy), and others with cold (think of ice). To control and prevent crystallization in the seal chamber it is necessary to control the temperature.
   - Suction Bypass
   - External Flush
   - Thermal Jacketed seal chamber
   - Quench and Drain
   - Heat Exchanger
   - Dual Seal
2. **Liquids that Solidify** To deal with liquids that try to solidify (think of ice, cement, glue and paints) it is necessary to identify the cause and solution for the solidification:

CAUSE OF SOLIDIFICATION	SOLUTION
1. Air (paint, glue)	Quench and Drain or the Double Seal
2. Change in Temperature (ice)	Control the temperature, either up or down. See Crystallization.
3. Agitation (merengue)	Evacuate the seal chamber, suction bypass. Vacíe la cámara del sello con una línea hasta la
4. Time (cement)	Evacuate the seal chamber, suction Bypass.

3. **Vaporizing Liquids** Certain liquids vaporize with heat (think of steam), and other liquids vaporize with a drop in pressure (think of liquid propane or freon). To control vaporizing liquids so they don't change phase in the seal chamber.

CAUSE OF VAPORIZATION	SOLUTION
1. Too much heat	Cool the seal chamber. See Crystallization.
2. Drop in Pressure	Double seal with pressurized barrier fluid

4. **Film or Skin Formation** Some liquids form a skin (like hot milk) with heat. Others form a film (like paints) on contact with air:

CAUSE OF FILM	SOLUTION
1. Temperature Control it	See Crystallization
2. Contact with air	Quench and Drain or Double Seal

5. **Very High Temperature** Remove the heat with cooling. See Crystallization.
6. **Non-Lubricating Liquids** No lubrication or heat dissipation. Remove the heat with a double seal and a barrier tank with forced convection flow.
7. **Dry Running Pump** Again, no lubrication or dissipation of heat. Remove the heat with a double seal and barrier tank with forced convective flow.
8. **Dangerous Liquids** Use a dual seal with a pressurized barrier tank.
9. **Gases and Liquids Tending to Gas** Gases cannot lubricate the seal faces. No dissipation of heat. Use a dual seal with forced convective flow.
10. **Slurries and Suspended Solids** Should be purged or flushed with:
    - Water
    - The filtered or cleaned product
    - A compatible liquid
    - A posterior additive in the system
11. **Cryogenics** Cryogenic liquids are too cold for most seals. Use a metal bellows mechanical seal with no elastomer parts.
12. **Corrosive Liquids** Vary with the temperature. Use seal components resistant to corrosion. You should take into consideration the:

- Elastomer parts and gaskets
- Springs
- Chemical compatibility with the face materials
- Metallurgy: sleeve, shaft, set screws, gland, metallic parts, drive pins, clips, keyways, anti rotation pins, etc.

13. **Very High Pressures** Use balanced o-ring cartridge seals up to about 500 psi. Above 500 psi use the tandem double seal with the barrier fluid pressurized at ½ the seal chamber pressure. Remember as pressure goes up the o-rings will extrude and metal parts will distort. Use a torsion balanced seal.
14. **Hard Vacuum** Use a balanced o-ring seal for industrial vacuum. For absolute pressure less than 1 kpa (1 kilopascal) use a torsion balanced seal. (Must verify this measurement)
15. **Extremely High Velocity Shaft Speed** Some pump companies use very high velocity, 30,000 rpm, to improve efficiency and generate high head with small equipment. Use a stationary seal, with the springs in the stationary element.

A big part of the overall problem with adequate mechanical seal life is trying to make a precision mechanical seal run into the same space that previously was occupied by the packing rings. Pump design evolved over time to accommodate the packing rings.

For example, the restriction bushing in the bottom of the packing stuffing box is designed to prevent the gland follower from pushing the packing out of the bottom of the box. With a mechanical seal, the restriction bushing in the bottom of the packing box is a 'dinosaur of design'. It performs no function with the mechanical seal, except to shorten its life by holding heat, preventing clean cooling liquid from arriving to the mechanical seal faces, and trapping abrasives, sediment, crystals, and dirt. If you were to remove the restriction bushing by placing the part on a lathe and machining it away, certain pumping applications could immediately quadruple the service life of the mechanical seal.

The bore of the packing box serves to hold the packing around the shaft so that the pressure from the gland follower can axially compress the rings to affect a shaft seal. With a mechanical seal, the reduced tight bore of the packing box is another dinosaur of design. It could be opened on a lathe in the same function as the removal of the restriction bushing at the bottom of the box. This would immediately triple or quadruple the service life of most mechanical seals.

Many packed pumps have an installed discharge bypass line running from the discharge nozzle of the pump to the packing box. This line damages most seals by blasting the seal with the highest concentration of solids moving through the pump. As a pump is converted from packings to mechanical seals, removing the discharge bypass line, opening the seal chamber bore, and machining the restriction bushing in the bottom of the chamber will go a long way to achieving the desired life with a mechanical seal.

Proper pump repair alignment methods

This dial indicator is fixed to the volute mounting adapter collar of the pump and the needle is on the shaft (Figure 14–21). The shaft should be moved radially by hand (see the arrows) up and down. Note the movement in the indicator. This is a check of the radial tolerance in the bearing. Some people use the word 'run out'. Radial deflection causes misalignment of the rotating and stationary faces of the mechanical seal. This shortens the seal life by causing drive pins and springs to wear and rub in relative motion.

Leaving the indicator needle touching the radial diameter of the shaft, rotate the shaft by hand (Figure 14–22). This should be done with the naked shaft, and also with the shaft sleeve if the pump takes a sleeve. This reading checks the roundness and straightness of the shaft. If the shaft is not round, there will be excessive vibrations and may distort the faces of the seal sacrificing it's optimum life. If the shaft is not straight, the rotary face of the seal will wipe across the stationary face as it spins. This leads to premature failure from vibrations and can damage the

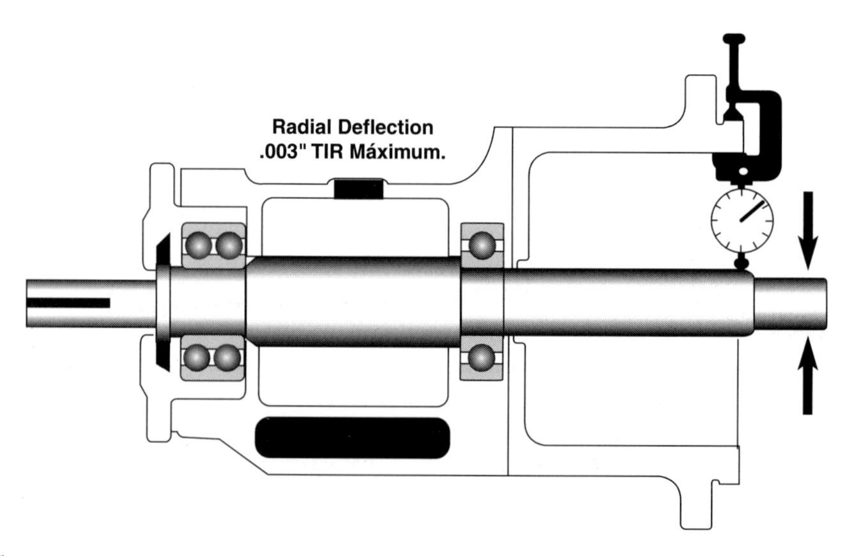

Figure 14–21

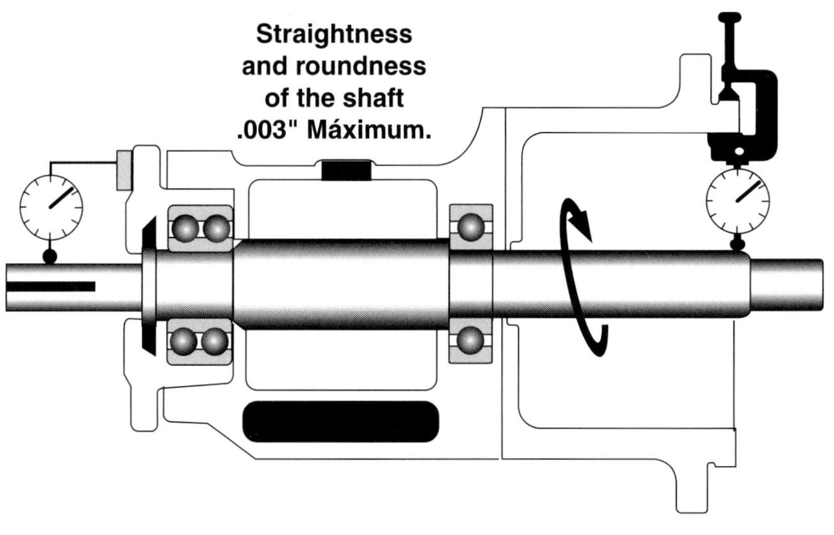

Figure 14–22

pump wear rings, and the motor and coupling. The same reading should be taken on the motor end of the shaft.

With the indicator fixed to the volute mounting collar, place the needle on the end of the shaft (Figure 14–23). Push and pull the shaft axially by hand (see the arrows). This will read the tolerance in the axial bearing. This tolerance affects the spring tension holding the faces together. If the play in the axial bearing is too loose, the movement can open or even crush the seal faces.

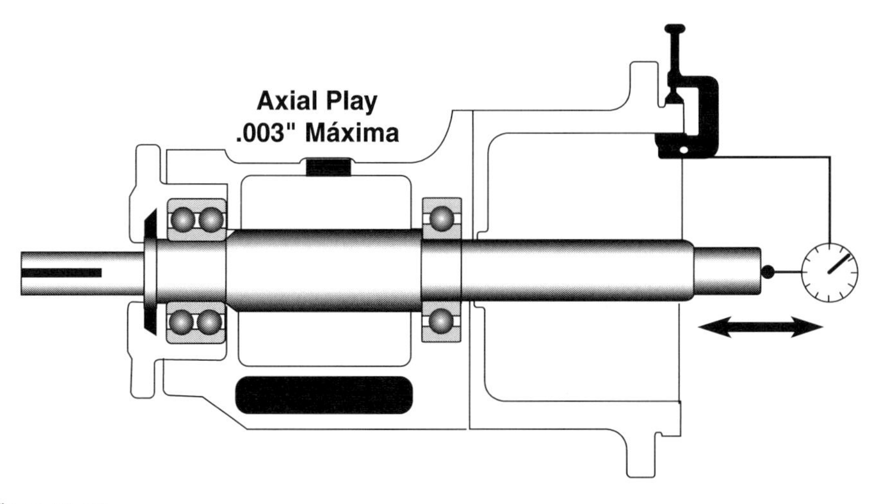

Figure 14–23

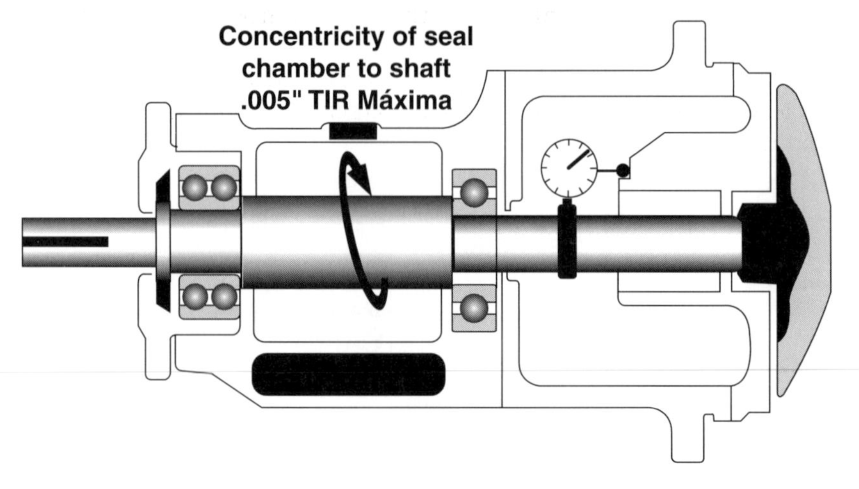

Figure 14–24

Install the pump back plate and seal chamber assembly. Mount the dial indicator on the shaft and place the needle onto the outer diameter of the lip or face of the seal chamber (Figure 14–24). An alternate method would be to place the indicator needle inside the seal chamber bore. Rotate the shaft. This will verify that the shaft is concentric with the seal chamber bore. If it is not concentric, the seal may rub against the bore when the pump is started.

With the indicator still in this same position, place the needle onto the lip or face of the seal chamber (Figure 14–25). Rotate the shaft. This

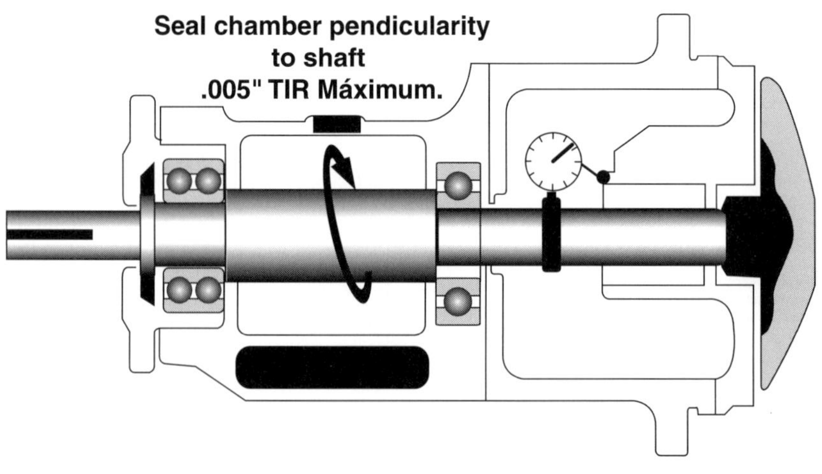

Figure 14–25

will verify the perpendicularity of the seal chamber to the shaft. If the chamber is not perpendicular to the shaft, the seal's faces and springs will have to flex twice with every revolution to maintain contact. This will lead to fretting corrosion, a damaged pump shaft or sleeve, and rapid failure of the seal.

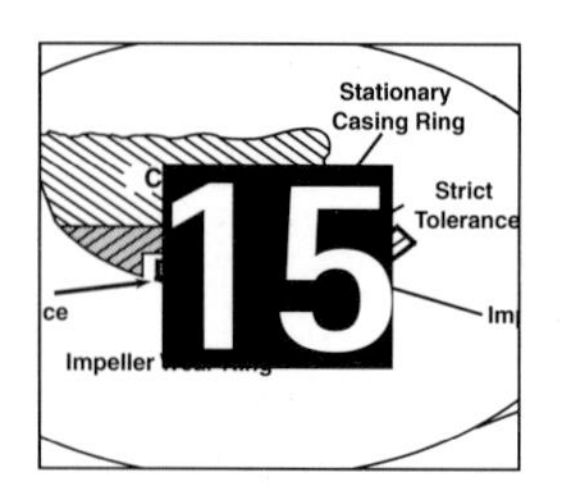

Common Sense Failure Analysis

Pump maintenance files

In most places, the available information about any pump, such as the manufacturer, year of purchase, model and serial number is placed in a file for general accounting purposes. In other plants, such as a manufacturer, the model, lubricant and lubrication frequency is placed in a lubrication schedule. In either case, additional key information can be stored with a little more effort.

The complete maintenance record of a pump, when filed in an accessible available place, is a valuable tool for diagnosing problems, ordering parts for repair, and establishing lubrication and maintenance schedules. Also, these maintenance files are valuable in determining the performance of the pump during process changes. The comments on the work orders, such as the list of materials or parts used, can define a good preventive frequency, a predictive and/or a planned maintenance repair. Equipment records with good information can help extend the period of inspection or identify specific checkpoints. In other pumps, the records can indicate frequent failures. These they can be classified as failures due to materials (incorrect parts), installation, maintenance and/or operation. Good and precise information in the record of the pump encourages applying a Root-Cause-Failure Analysis method. The result of this analysis can suggest more inspections and repairs as well as changes in operation procedures, frequency of lubrication or better inspection procedures from project inspectors when accepting a new installation. Organizing the data in chronological format is useful to diagnose problems, visualize what happened before it broke down and who carried out the repair. Keeping simple and complete maintenance records of each pump is more economical than trying to solve problems without information using the method of trial and error.

With an appropriate record of repairs, you can use this information to develop a correct parts inventory that is based on actual parts consumption and not on recommended parts provided by the manufacturer. The frequent replacement of worn parts can indicate a possible substitution of materials from the original OEM part.

Record keeping is critical in those industries whose production requires the use of many pumps. The record of the pump should have the complete information on the installation, application and maintenance. Space should be provided in each card, using both sides, to keep a complete record during a two-year period, and in some cases, for the whole life of the pump.

Failure analysis on centrifugal pumps

Many times, the broken part of a pump is replaced when it fails without an effort to understand why the situation happened. Any corrective action that takes place is usually a temporary arrangement. The probability is quite high that the pump will fail again for the same reason. This part replacement with no analysis practice is not acceptable due to the high cost of the maintenance, parts, time and lost production.

It is interesting to note that some pump users literally know that their pumps will fail after a specific time period. They understand that the running time of the pump should be maximized to have an acceptable yield in the process. This type of strategy is expensive since it raises a doubt of the continuity of the pump performance. To compensate, some plants install back up or redundant pumps.

In order to solve a pump failure, we have to identify the cause. Once this is known, the problem can be dealt with and a permanent solution can be found. A logical thought process (common sense) to identify the problem is as follows:

1. Ask 'What's making this happen?' – It is likely that what we call the problem is actually the symptom. Example: 'Low discharge pressure', 'failed mechanical seal', 'the pump makes noise.'
2. Look for the evidence – The evidence is the manifestation of the symptoms. The evidence indicates that there is a problem with the pumping system. Example: 'the discharge gauges indicate a low pressure'.
3. Verify evidence – Example: 'Is the gauge calibrated and accurate?' Eliminate or cancel other reasons or possibilities for the evidence. Example: 'The pump is not pumping enough pressure and we're no longer able to fill that tank.'

4. Identify the causes supporting the evidence. Example: What could cause low pressure? The cause is the origin of the failure.

The causes of low pressure, for example, could be either hydraulic or mechanical. In many cases of failure analysis, asking 'Why?' and 'What?' and answering those questions, until you can no longer ask 'why', will almost always get you to the answer. If all evidence leads to a mechanical reason for the failure, the problem is probably maintenance induced. If the evidence leads to a hydraulic reason for the failure, the problem is either operations or design induced. In cases where the 'reason for failure' was not determined, a more extensive analysis is necessary. The additional analysis is recommended to take advantage of the pump supplier experience in identifying the root cause.

CASE STUDY

A paper mill was using an ANSI end suction process pump with clear water service. The motor was designed properly. The pump axial thrust bearing ran hot, failing after three months of operation. It was replaced with an identical bearing. This ran during three months and also failed. All pump components were investigated and found that they complied with the specifications. These facts eliminated the defects of materials as a cause.

All the failed parts in the unit were inspected to assure that they were manufactured according to specifications. This step eliminated defects from the factory.

The maintenance team, that is to say, the mechanics, were found competent and they followed the correct maintenance procedures. This information eliminates the defects by maintenance.

Although the pump was being run at 25% of the BEP (Best Efficiency Point), it was designed for this type of service. This eliminated improper operation or running outside of the Sweet Zone.

There remains only one cause to explore, the design of the pump. The manufacturer was contacted and their design group studied the situation. By opening balance holes on the face of the impeller, they reduced the heat generated by 70°F. The manufacturer also incorporated a flinger ring lubrication system, a bigger oil reservoir to cool the oil and improved the oil circulation. The plant has not experienced a failure of the pump since the redesigned pump was installed.

Why is this pump in the shop?

Did you ever notice that the building or area in the plant called the 'maintenance shop', is actually the Pump Hospital? The shop may have twelve workbenches, but ten benches have a pump in some stage of surgery. You go into the shop and ask someone 'Why is this pump in the shop?' And someone says, 'Because it was making noise', or 'The seal failed'.

The noise and the seal failure are actually symptoms and not the problem. This is like the electrician blaming the fuse for an overloaded electrical circuit. The problem is the overloaded circuit and the symptom is the burned fuse. Likewise, in the maintenance shop, the noisy pump, the failed seals and bearings are the 'Symptom' of a problem that probably occurred outside the pump.

In this book, we've dedicated whole chapters to seals and bearings. However, there are some other complaints (symptoms) that send pumps into the shop. We have listed below some of those reasons. We present them in table form with the symptom and the possible hydraulic and/or mechanical cause for the symptom. We hope this helps someone.

SYMPTOMS AND POSSIBLE ROOT-CAUSES

Symptom	Possible Hydraulic Cause	Possible Mechanical Cause
Noisy Pump.	Cavitation Aspirated Air Excessive Suction Lift Not enough NPSHa	Bent Shaft Bound Rotor Worn Bearings
Not enough discharge flow	Excessive discharge Head Not enough NPSHa	Worn or damaged impeller Inadequate foot valve size. Air aspiration or air pocket in the suction line. Plugged impeller or piping
No discharge pressure.	Pump improperly primed. Inadequate Speed. Not enough NPSHa.	Plugged impeller or piping. Incorrect rotation. Closed discharge valve Air aspirated or air pockets at the suction line.
Pressure Surge.	Not enough NPSHa.	Air aspirated or air pockets at the suction line. Entrained Air. Plugged impeller.
Inadequate Pressure.	Not enough velocity. Air or gases in pumped liquid.	Impeller diameter too small Worn or damaged impeller Incorrect rotation
Excessive Power Consumption	Head too small, excess flow. High specific gravity or high viscosity.	Bent shaft. Bound shaft. Incorrect rotation.

Although about half of all pumps manufactured in the world are centrifugal (the other half are positive displacement), industry tends to use a higher quantity of centrifugal pumps. For that reason, much of this book has dealt with pump theory, applications, and problems, from a centrifugal point of view. You may think that we have abandoned PD pumps in this book. You would be wrong.

Actually, everything we said about bearings, mechanical seals, piping, TDH, system curves and mating the pump curve to the system curve, the affinity laws, cavitation, horsepower and efficiency are as applicable to PD pumps as centrifugal pumps.

So in this chapter of failure analysis and corrective methods, we decided to consider some problems, symptoms, and remedies particular to PD pumps. We're using two tables. The first table lists the few symptoms that send a PD pump into the shop. These symptoms are mated to another column of possible causes listed in numerical order. The numerical causes are on the second table starting with the source of the problem in the left column and the probable cause/suggested remedy in the right column. As you go through the list, you'll see again that PD pumps and centrifugal pumps have a lot in common. Enjoy.

SYMPTOMS AND CAUSES OF FAILURE FOR POSITIVE DISPLACEMENT PUMPS

Symptom	Possible Cause
Pump fails to discharge liquid.	1,2,3,4,5,6,8,9
Noisy pump.	6,10,11,16,17,18,19
Pump wears rapidly.	11,12,13,16,20,23
Pump not up to capacity.	3,5,6,7,9,21,22
Pump starts, then loses suction.	1,2,6,7,10
Pump consumes excessive power.	14,16,17,20

Source of Problem	Suggested Cause/Remedy
1. Suction problem.	Not properly primed
2. Suction problem	Suction pipe not submerged
3. Suction problem.	Clogged strainer
4. Suction problem.	Foot valve leaks
5. Suction problem.	Suction lift too high
6. Suction problem.	Air leak in suction piping
7. Suction problem.	Suction piping too small

	Source of Problem	Suggested Cause/Remedy
8.	System problem.	Wrong rotation
9.	System problem.	Low speed
10.	System problem.	Insufficient liquid supply
11.	System problem.	Excessive discharge pressure/resistance
12.	System problem.	Grit or dirt in liquid.
13.	System problem.	Pump running dry
14.	System problem.	Viscosity of liquid being pumped is higher than specified.
15.	System problem.	Obstruction in the discharge line
16.	Mechanical problems.	Unbalanced or misaligned coupling.
17.	Mechanical problems.	Bent motor shaft
18.	Mechanical problems.	Chattering relief valve
19.	Mechanical problems.	Pipe strain distorting the pump casing.
20.	Mechanical problems.	Air aspiration thru the packing/seal.
21.	Mechanical problems.	Inadequate relief valve.
22.	Mechanical problems.	Packing is too tight.
23.	Mechanical problems.	Corrosion.

AUTHOR'S NOTE

Oh yes, 'Common Sense Maintenance' is likely to be the title of our next book.

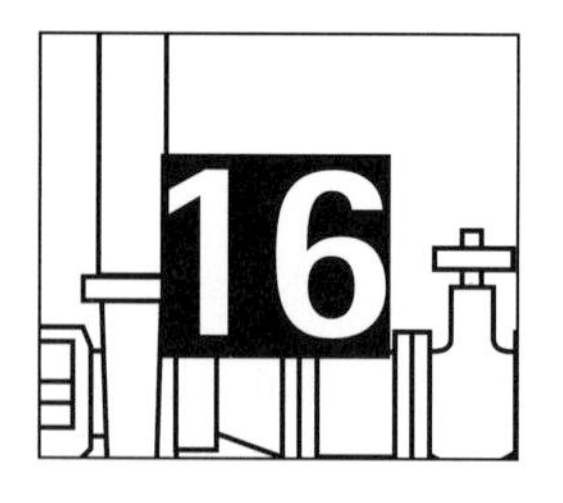

Avoiding Wear in Centrifugal Pumps

Introduction

In the moment of starting a new pump, that pump is headed for the day when it will need repair even if the design and operation is correct. One factor that determines the repair is internal wear. Imagine an ideal application where the pump is operating at its BEP and the system is stable. Does this condition ever exit? If you answer yes, you are one of fortunate few. However at some point, even if it does not break, the pump will go to the shop because of internal wear. This chapter presents different sources of internal wear and suggestions to extend the useful running time of the pump.

Erosion

Erosion is the wear of the pump internal parts by suspended solid particles contained in the fluid being pumped. The most affected parts are: wear rings, shaft sleeves, packing, mechanical seal faces, lip seals, the pump casing and the impeller.

Erosion can be caused by small particles not visible to the human eye, like dissolved minerals in 'hard water.' Larger solids like sand, boiler scale, and rust can also cause serious erosion inside the pump.

The fluid being pumped is often not well defined. Terminology like well water, industrial effluent, raw water, boiler feed water, condensate water, etc., is usually the only definition we have of the fluid being pumped. Any of these fluids can contain several concentrations of solids that cause erosion and wear inside the pump.

When the liquid being pumped is known to have a large concentration of solids, the materials inside the pump should be changed to more resistant materials. Materials such as carbon steel, high chrome iron, harden stainless steel or hard coatings like ceramic or tungsten alloy are some of the most used.

Corrosion

Corrosion is caused by a chemical or electrochemical attack on the surface of the metals. It is increased when there is an increase in temperature and/or presence of oxygen in the fluid or the surface of the fluid.

We can aggravate the corrosion effect if misaligned parts have relative movement, such as loose fit bearings or rapid changes in the system. Cavitation, erosion and high fluid velocity advance the corrosion process.

Cast Iron is a widely used material for centrifugal pump housings. It is used when the fluid PH is 6 or higher (not acid). Cast Iron corrodes and forms a protective coating on the surface of the metal. This graphitized surface protects the metal from further corrosion as long as the coating is intact. High velocity fluids, cavitation, metal to metal contact and erosion can affect this protective coating.

If corrosion exists, the pump-wet parts can be changed for other materials such as stainless steel or composite material. Impellers can be replaced by bronze cast impellers or other materials.

Wear rings

Wear rings provide for a close running, renewable clearance, which reduces the amount of liquid leaking from the high pressure zones to the low pressure zones in the pump. They are commonly fitted in the pump casing and on the impeller (Figure 16–1, next page).

These wear rings are lubricated with the fluid being pumped. Eventually they will wear. Tolerances open and more liquid passes from the discharge end back to the suction end of the pump. The rate of wear is a function of the pumped liquid's lubricity. When the wear is excessive, the pump suffers degradation in its performance. This is particularly true with small pumps running at high speed. The strict tolerance in the replaceable wear rings governs the efficiency of the pump. When the pump goes to the shop, these wear rings should be changed.

You can expect the pump to loose 1.5 to 2% efficiency points for each one thousandths (0.001 inch) wear in a wear ring beyond the original

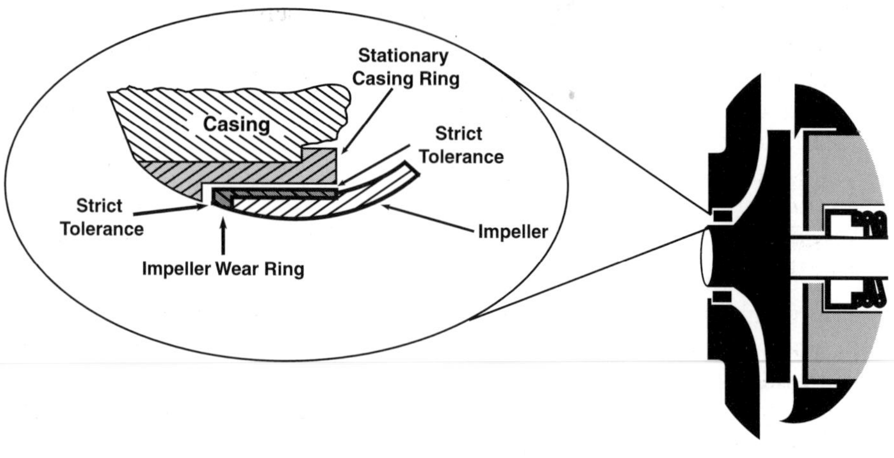

Figure 16-1

factory setting. This setting is based on the operating temperature of the application. Let's consider how much money the lost efficiency costs. We will use some formulas from Chapter 5 in this book on useful work and efficiency.

$$\textbf{Cost per year} = \frac{\textbf{0.000189} \times \textbf{GPM} \times \textbf{TDH} \times \textbf{\$Kwh} \times \textbf{sp.gr.} \times \textbf{8,760}}{\textbf{Eff. Pump} \times \textbf{Eff. Motor}}$$

where: 0.000189 = Conversion factor
GPM = Gallons per Minute
TDH = Total Dynamic Head
$Kwh = Cost per Kilowatt-hour
sp.gr. = Specific Gravity
8,760 = Hours in a year
Eff. Pump = Pump Efficiency
Eff. Motor = Motor Efficiency

To show cost increase, consider this newly installed pump in a properly designed system. We have the following values:

GPM = 2,000 gpm
TDH = 120 ft.
$Kwh = $ 0.10
sp.gr. = 1 (water)
Eff. Pump = 77%
Eff. Motor = 93%

The electricity cost to run this pump for a year is $55,450.80.

After being in line for six months, this pump is disassembled and it is noted that the tolerance in the wear bands has opened 0.004 inch from the original factory setting. This wear represents an 8% decrease in efficiency. Now the pump is 69% efficient. Let's do the math with all other factors constant. This reduction in the efficiency represents an annual electricity cost of $61,845.60. The additional electricity is six thousand three hundred ninety four dollars and eighty cents. Four thousandths wear (0.004 inch) has cost us almost $6,500.00 per year for just one pump. Just to mention, a new wear ring may cost up to $60.00 plus the labor to change it (this will never add up to $6,500.00).

Effective and well planned maintenance can reduce the operating cost of your pumps and other equipment as this example demonstrates. With differential pressure gauges on the pump, an amp meter and flow meter you can determine if strict tolerance parts are worn. This indicates the need to take the pump into the shop for corrective procedures. If you don't do it, you are wasting your annual operating budget. As we mentioned in Chapter 6, the Wear Rings should be called Efficiency Rings. Now you know why.

Fluid velocity accelerates wear

Small impellers with high motor speeds may produce the necessary pump pressure. This type of combination produces high fluid velocities that will wear pump parts much faster than desirable. This is in the Affinity Laws. In addition the impeller suffers rapid wear due to high tip velocities. When a pump is disassembled and excessive wear is found, 95% of the time high velocity fluid is to blame.

Turbulence

Uneven wear in parts is often due to turbulence. Bad piping designs or poorly sized valves can cause turbulence and uneven wear in pumps. Whenever possible, use straight pipe sections before and after the pump. Uneven flow creates turbulent flow and excessive wear occurs.

It is not recommended to place an elbow at the suction of any pump (Figure 16–2, next page). This will cause a turbulent flow into the pump. If elbows are needed on both sides of the pump, you should use long radius elbows with flow straighteners. You should have 10 pipes diameters before the first elbow on the suction piping (Example: If the pump has a 4 inch suction nozzle, you should respect 40 inch of straight pipe before the first suction elbow.) Short radius elbows cause vibrations and pressure imbalances that to lead to wear and maintenance on the pump.

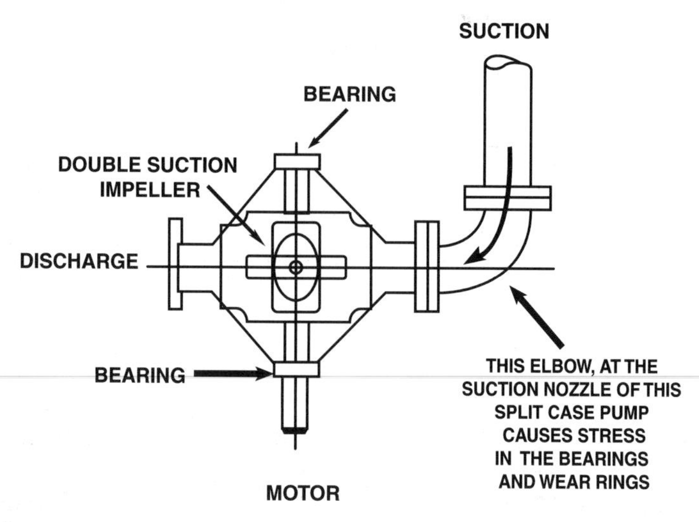

Figure 16–2

A pipe size increaser can be used in the discharge piping. This will reduce the fluid velocity and friction losses. An isolation valve with a low loss characteristic such as a gate valve should be placed after the increaser and check valve.

Throttling

A centrifugal pump should never be operated continuously at or near the shut off head. This normally happens when a tank or vessel is near the maximum capacity and an operator or level sensor starts closing the discharge valve while the pump is running. This is similar to activating your car brakes while the gas pedal is to the metal. All this wasted energy is transferred to the fluid being pumped. This type of operation shortens the life of the pump and increases the downtime. This energy is converted into heat and vibration raising the fluid temperature. Some pump casings can dissipate the heat. Other casings contain heat switches that will trip-out and 'shut off' the pump.

An intensive radial load is created when operating near the shut-off head and the shaft deflects at about 60° from the cut-water. This concept is explained in Chapter 9 'Shaft Deflection'. The pump will be noisy, will vibrate and maintenance on seals, bearings and shaft sleeves is expected.

Pumps are usually over-designed. From the initial specification stages, future needs are taken into consideration, maximum flow is overrated and operating conditions are uncertain. Design engineers following a

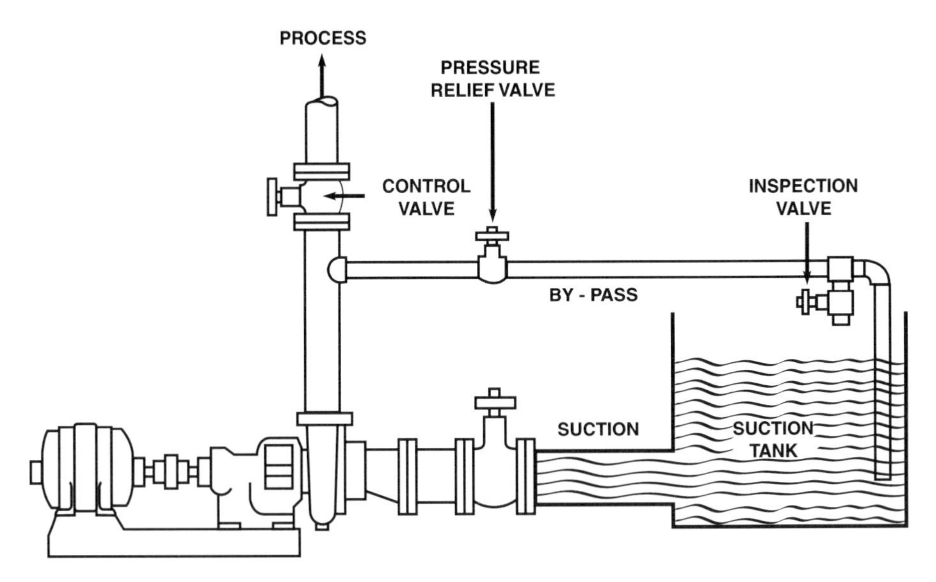

Figure 16–3

financial guide to lower future capital investment do this. With this in mind, no wonder we have to extensively throttle the discharge in order to achieve our necessary flow rate. Yes, this saves future capital investments, but creates present problems with the daily maintenance budget. Excessive throttling of the discharge valve results in severe punishment to the pump.

Consider Figure 16–3. This pump is draining a tank and discharging into the process stream. If a control valve should be strangled to a certain predetermined pressure (resistance), the pressure relief valve automatically opens and the excess discharge pressure recirculates back to the suction tank. The pump never knows that the discharge control valve is throttled. If this situation exists in your plant or if the operators regulate the flow by manipulating control valves, this will be the proper design to extend the useful life of your pump.

Pump Piping

Introduction

We all know that piping is integral to the pump system. Because it is connected to the suction and discharge, the piping affects the health and well being of the pump. Incorrect pipe installation prejudices the pump's useful life.

In this chapter, we present graphic information on inadequate and correct piping arrangements.

Piping design to drain tanks and sumps

When draining a tank with two pumps, you should not use a 'T' with two connections. The dominant pump may asphyxiate the other pump. Each pump needs its own supply pipe (Figure 17–1).

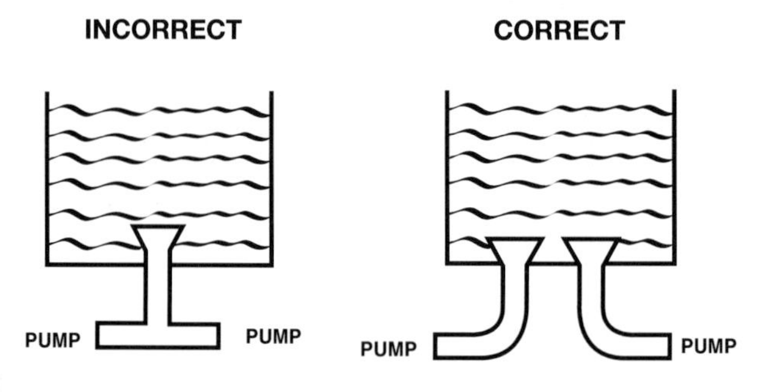

Figure 17–1

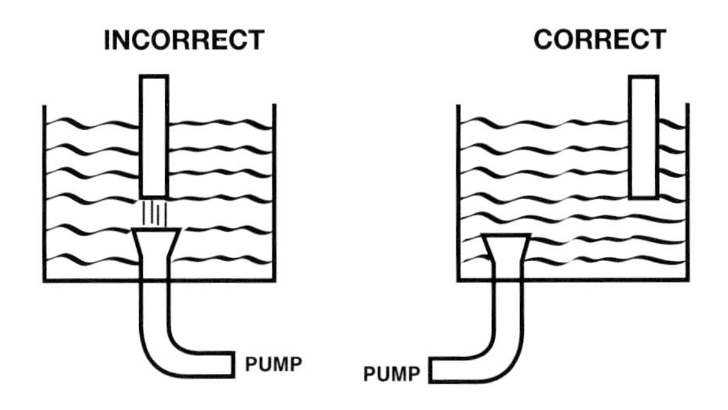

Figure 17–2

The in-flow pipe should not cause interference with the drain pipe (Figure 17–2).

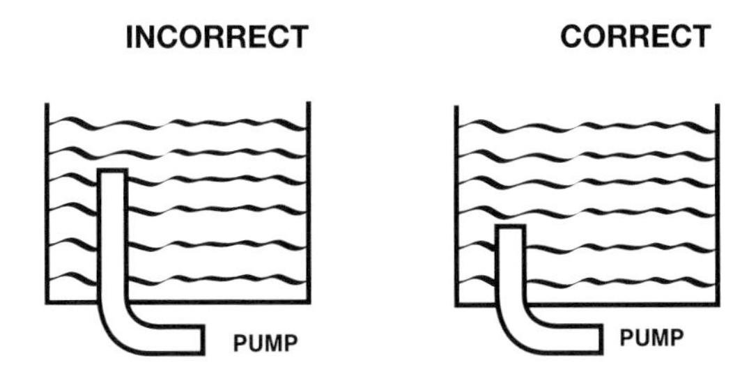

Figure 17–3

Drain pipe design must respect proper submergence (Figure 17–3). The submergence laws appear later in this chapter.

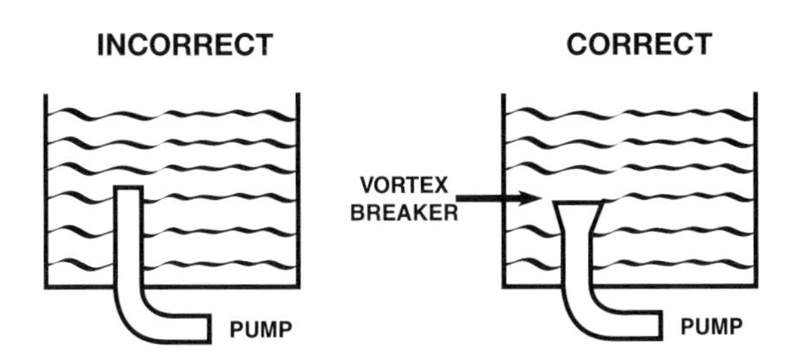

Figure 17–4

Use vortex breakers (Figure 17–4).

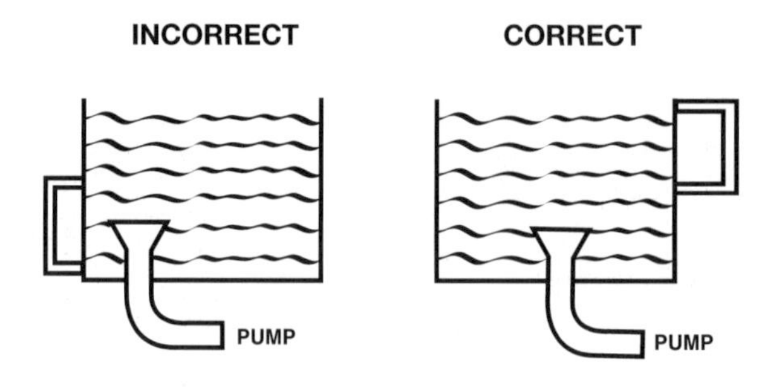

Figure 17–5

Design the level indicators to respect the proper submergence (Figure 17–5).

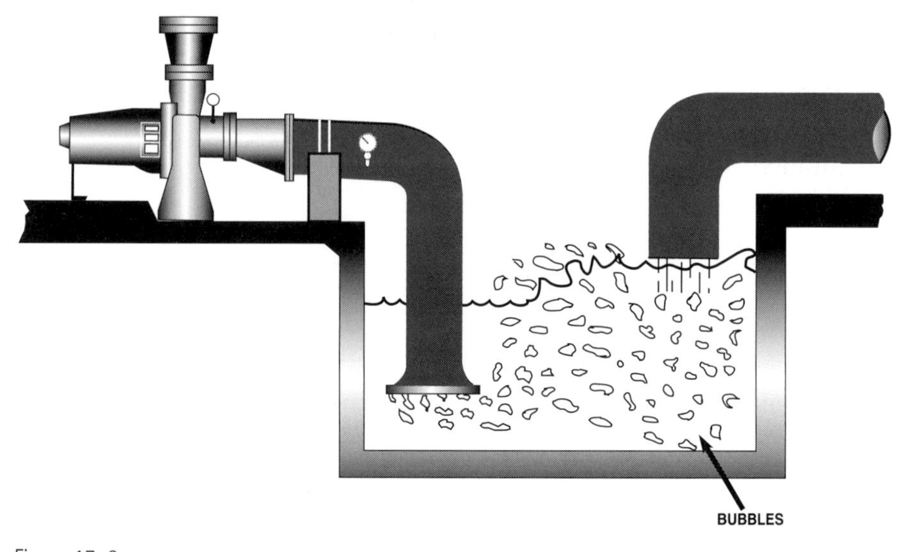

Figure 17–6

Inadequate sump design leads to entrained air bubbles and turbulence. This will damage the pump (Figure 17–6).

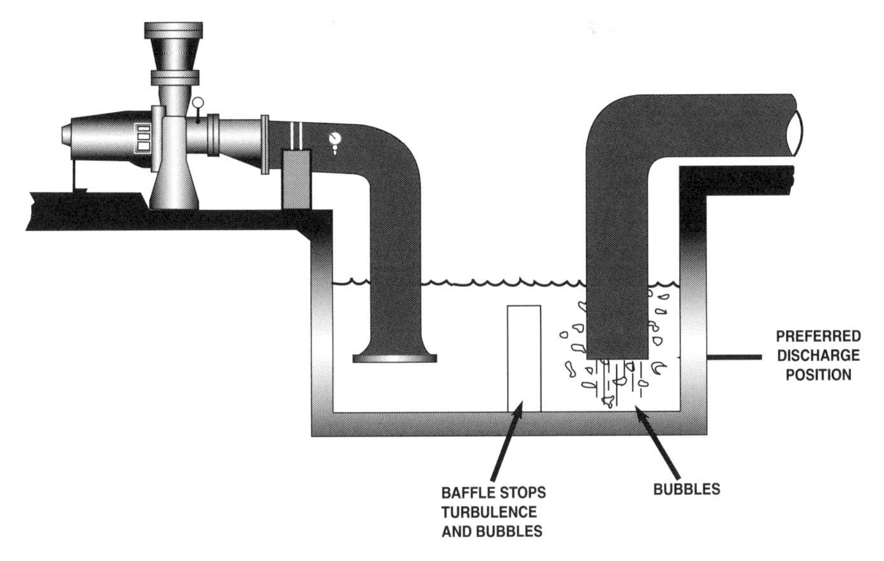

Figure 17–7

A submerged in-flow pipe and tank baffles prevent turbulence and bubbles from entering the suction piping (Figure 17–7).

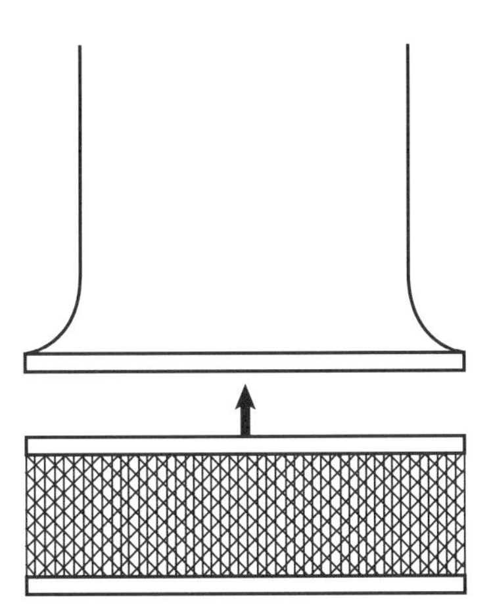

Figure 17–8

The suction bell reduces entrance losses and helps to prevent vortices. If you use a basket strainer, the screen area should be four times the area of the entrance pipe. Avoid tight mesh screens because they clog quickly (Figure 17–8).

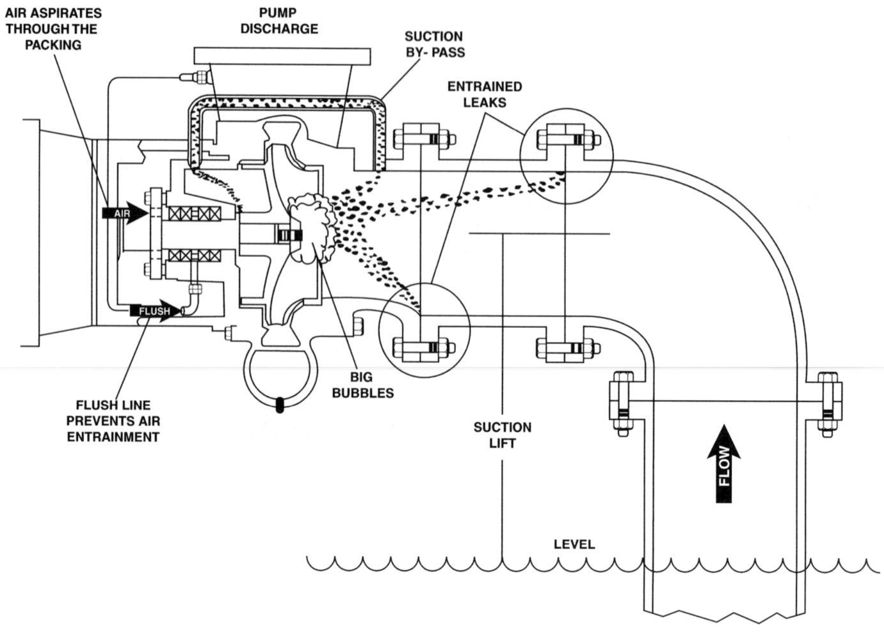

Figure 17-9

Avoid high speed suction flow (Figure 17–9). This causes air entrainment. Also, a high suction lift produces the same effect.

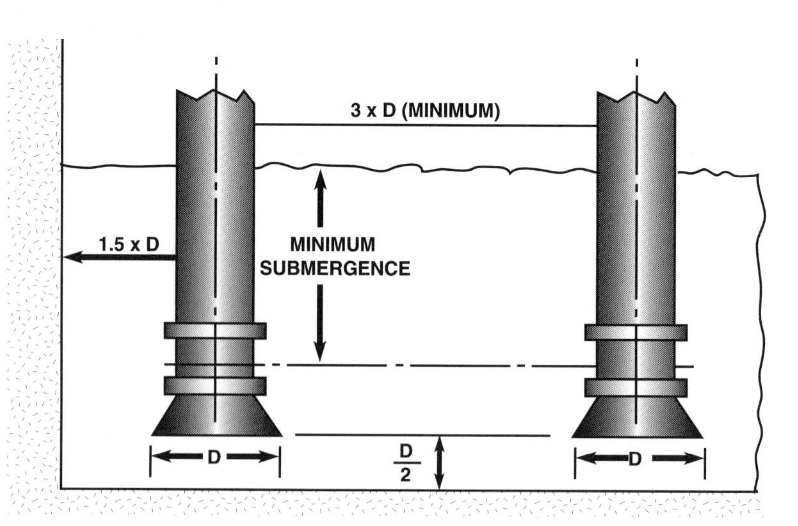

Figure 17-10

These are the dimensions to respect for proper sump design (Figure 17–10). The submergence laws are independent of the pumps NPSHr. The submergence laws are presented later in this chapter.

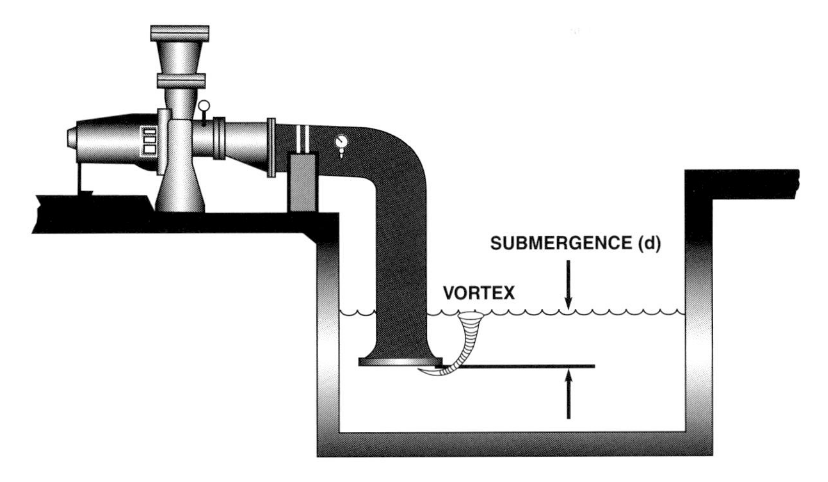

Figure 17–11

This aspirated air vortex is the result of not respecting adequate submergence (Figure 17–11). The submergence laws follow.

The Submergence Laws

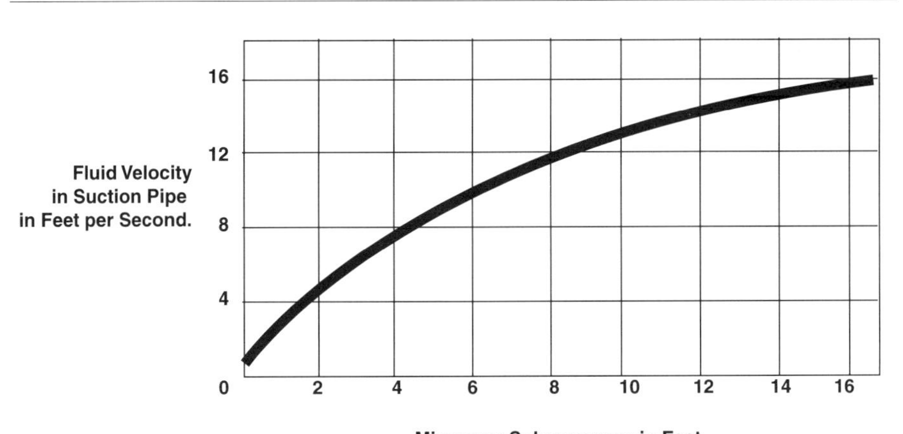

Figure 17–12

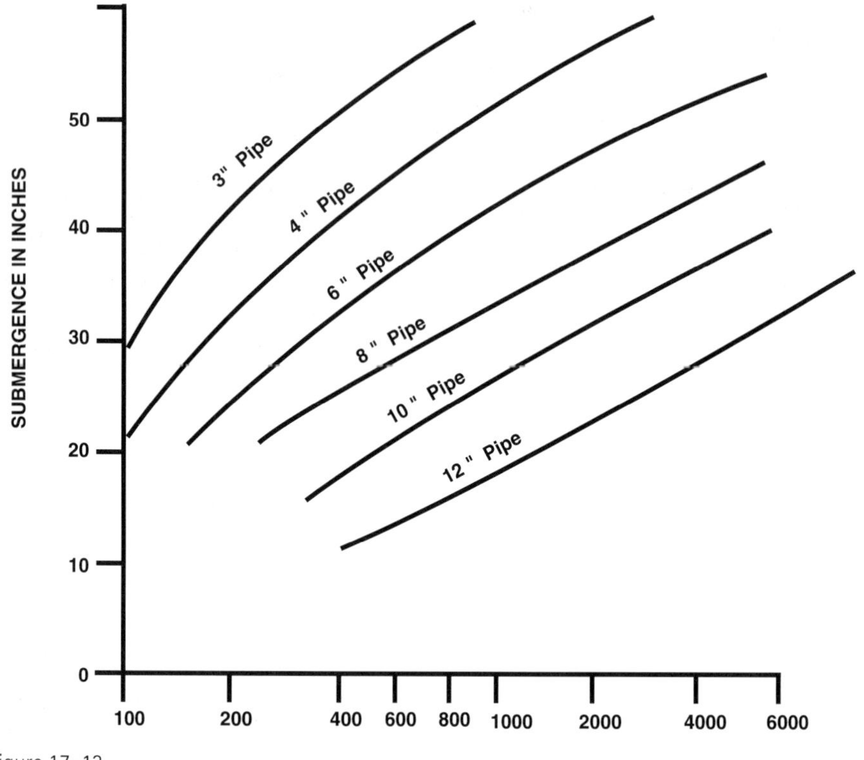

Figure 17-13

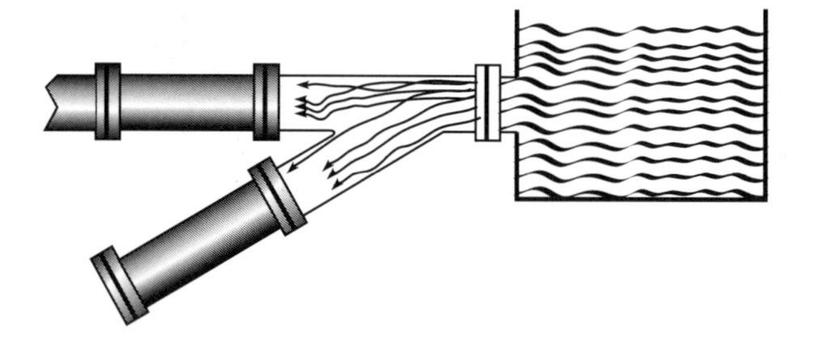

Figure 17-14

Use a 'Y' branch and not a 'T' branch. This will reduce turbulence.

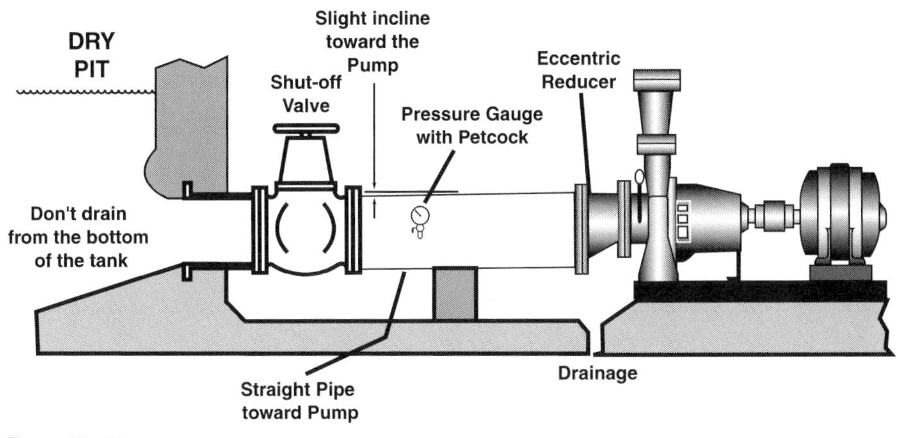

Figure 17–15

Correct suction piping leading to the pump (Figure 17–15).

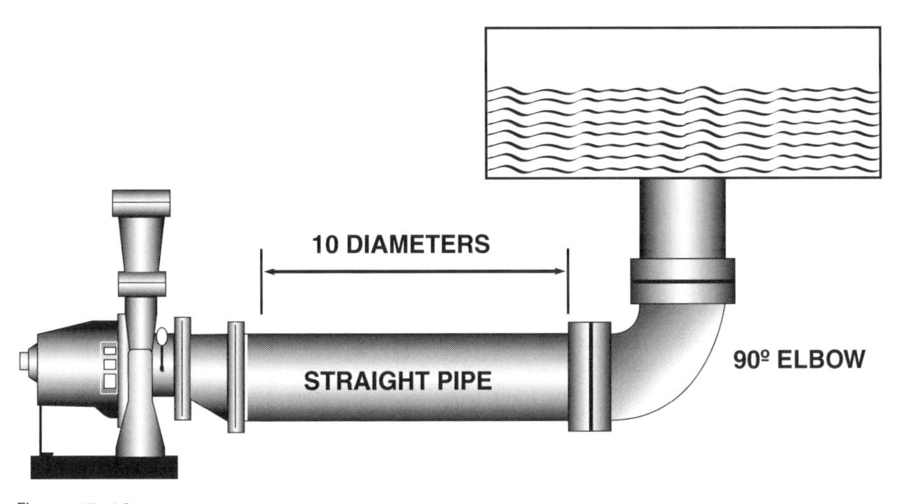

Figure 17–16

You should respect 10 pipe diameters before the first elbow in the suction piping (Figure 17–16). Example: If the pump has a 6 inch suction nozzle, you should have 60 inches of straight pipe before the first elbow.

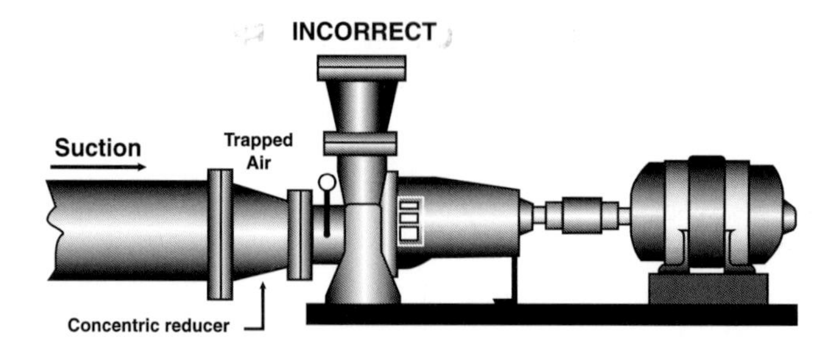

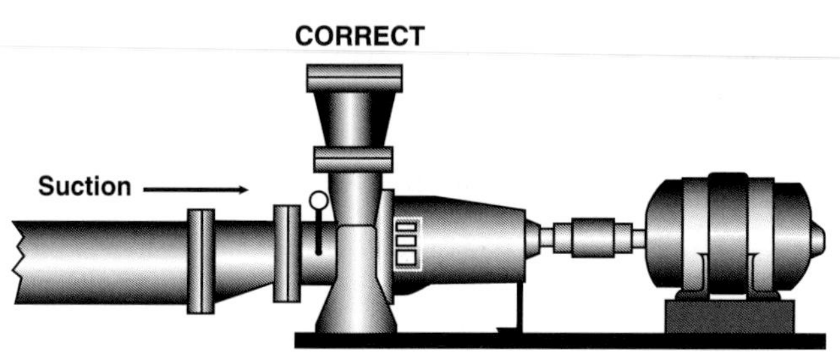

Figure 17–17

Use an eccentric pipe reducer to connect to the pump suction nozzle (Figure 17–17).

Don't use flange bolts to unite misaligned piping to the pump. This damages the flange faces and stresses the pump casing (Figure 17–18).

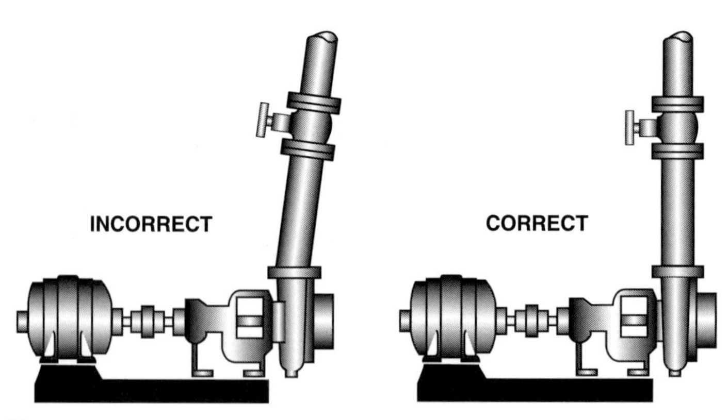

Figure 17–18

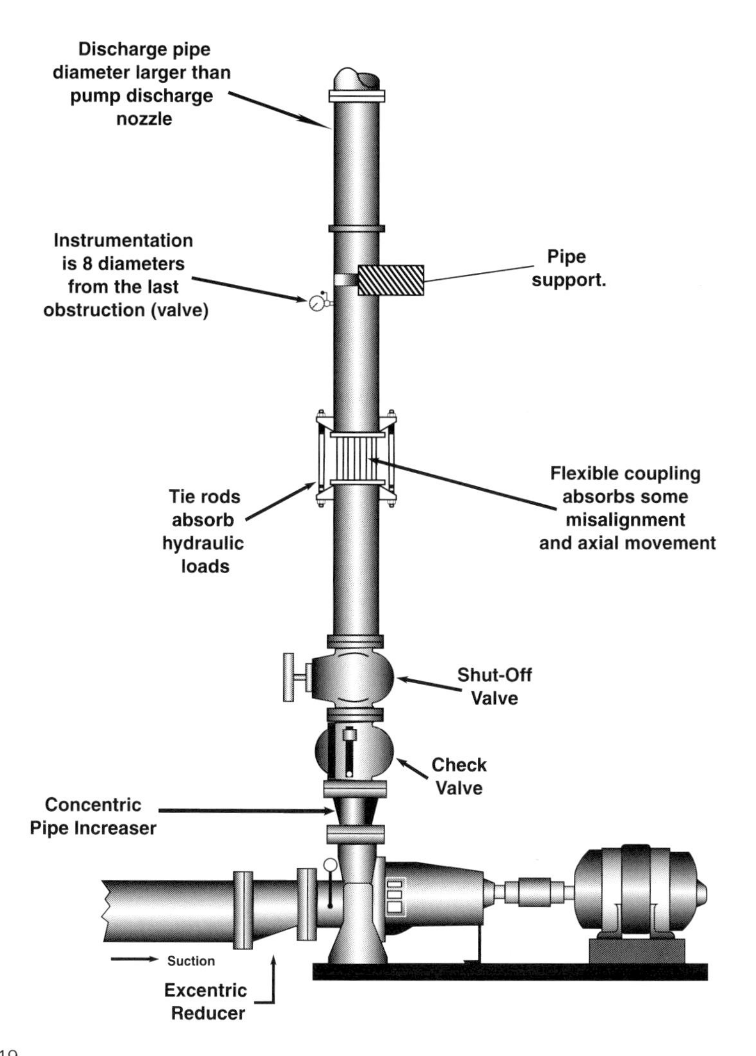
CORRECT DISCHARGE
PIPING
Discharge pipe
diameter larger than
pump discharge
nozzle
Instrumentation
is 8 diameters
from the last
obstruction (valve)
Pipe
support.
Tie rods
absorb
hydraulic
loads
Flexible coupling
absorbs some
misalignment
and axial movement
Shut-Off
Valve
Check
Valve
Concentric
Pipe Increaser
Suction
Excentric
Reducer

Figure 17-19

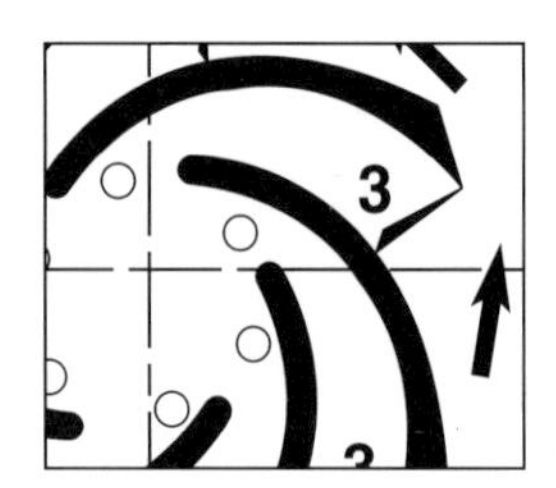

Index

RECEIVED
OCT 10 2003